RABID

ALSO BY BILL WASIK

And Then There's This:
How Stories Live and Die in Viral Culture

Frontispiece of a book about a plague in Venice, 1657.

RABID

A CULTURAL HISTORY OF THE WORLD'S MOST DIABOLICAL VIRUS

BILL WASIK AND
MONICA MURPHY

VIKING

VIKING
Published by the Penguin Group
Penguin Group (USA) Inc., 375 Hudson Street, New York, New York 10014, U.S.A. • Penguin Group (Canada), 90 Eglinton Avenue East, Suite 700, Toronto, Ontario, Canada M4P 2Y3 (a division of Pearson Penguin Canada Inc.) • Penguin Books Ltd, 80 Strand, London WC2R 0RL, England • Penguin Ireland, 25 St. Stephen's Green, Dublin 2, Ireland (a division of Penguin Books Ltd) • Penguin Books Australia Ltd, 250 Camberwell Road, Camberwell, Victoria 3124, Australia (a division of Pearson Australia Group Pty Ltd) • Penguin Books India Pvt Ltd, 11 Community Centre, Panchsheel Park, New Delhi–110 017, India • Penguin Group (NZ), 67 Apollo Drive, Rosedale, Auckland 0632, New Zealand (a division of Pearson New Zealand Ltd) • Penguin Books (South Africa) (Pty) Ltd, 24 Sturdee Avenue, Rosebank, Johannesburg 2196, South Africa

Penguin Books Ltd, Registered Offices:
80 Strand, London WC2R 0RL, England

First published in 2012 by Viking Penguin,
a member of Penguin Group (USA) Inc.

10 9 8 7 6 5 4 3 2 1

Image credits
Pages iv and xii: Wellcome Library, London; 14: © Leeds Museums and Art Galleries (City Museum), UK / The Bridgeman Art Library International; 38 and 202: Photograph by the authors; 64: © The Trustees of the British Museum; 90: Private Collection / Peter Newark American Pictures / The Bridgeman Art Library; 118: Bibliothèque des Arts décoratifs, Paris / Archives Charmet / The Bridgeman Art Library; 150: Penguin Group (USA) Inc.; 180: *Annals of Internal Medicine;* 224: Digital image © 2011 Museum Associates / LACMA. Licensed by Art Resource, NY.

Library of Congress Cataloging-in-Publication Data

Wasik, Bill.
Rabid : a cultural history of the world's most diabolical virus / Bill Wasik and Monica Murphy.
p. cm.
Includes index.
ISBN 978-0-670-02373-8
1. Rabies—Epidemiology—History. 2. Rabies—Treatment—History. I. Murphy, Monica. II. Title.
RC148.W37 2012
614.5'63—dc23 2011043903

Printed in the United States of America
Set in Haarlemmer with Engravers
Designed by Daniel Lagin

For our "creatures"—Emmett and Mia

CONTENTS

RABID

From a Spanish edition of Dioscorides' *De Materia Medica*, 1566.

INTRODUCTION

LOOKING THE DEVIL IN THE EYE

Ours is a domesticated age. As human civilization has spread itself out over the march of millennia, displacing wildlife as we go, we have found it advisable to strip the animal kingdom of its armies, to decommission its officers. Some of these erstwhile adversaries we have hunted to extinction, or nearly so. Others we relegate to zoos, confine in child-friendly safari parks. The balance we shunt to the margins as we clear the land for ourselves—erecting our own sprawling habitats on the ruins of theirs, naming our cul-de-sacs for whatever wilderness we dozed to pave them.

Peer through news reports, though, and one can find pockets of resistance, as if some ancient animal instinct were furtively reasserting itself. Consider the kamikaze bobcat in Cottonwood, Arizona, that set out on a rampage one recent March evening, menacing a worker outside a Pizza Hut and then sauntering into a bar, sending patrons onto the pool table, mauling the one who dared to snap a picture on his phone. Or the frenzied otter in Vero Beach, Florida, at a waterfront golf community called Grand Harbor, that gnawed three residents, one of them while out on the links. Or the enraged beaver in the Loch Raven Reservoir, in the genteel exurban sprawl north of Baltimore, that cruelly interrupted the summertime reverie of four swimmers; it was deterred from devouring them only when the husband of one victim pulled it from her thigh and smashed it with a large rock.

Typically, such wildlife shuns the society of humans. But in an instant we can find these meek woodland creatures transformed into bewilderingly avid attackers, accosting us as we retrieve our mail or walk our dogs. Sometimes they will even attempt a home invasion, as one young couple in the Adirondack hamlet of Lake George, New York, learned on an April evening just a few years ago. Walking from their car, they were set upon by a gray fox; they managed to rush inside their home and close the door. But nearly a half hour later, when they opened the door again, the fox lay in wait; it sprinted toward the opening. Only quick reflexes allowed the young man to close it just as the devil's snout broached the threshold. When an animal control officer arrived, the fox attacked his SUV, repeatedly sinking its teeth into the vehicle's tires. He shot at it multiple times from out his driver's side window but failed to hit his mark. Later, after the officer had finally run the fox over, he told a reporter that it was the single most aggressive foe he had encountered in nine years on the job. "This was a four- or five-pound animal attacking a 3,000-pound vehicle," he said.

The sheer tenacity: that is the truly chilling element in all these tales. "What disturbs me," remarked one Connecticut man to the local news, regarding the raccoon he had lately beaten to death with a hammer, "is I smashed his mouth off, I smashed his teeth in, but he still wanted to continue in the attack mode. I was actually terrified at the resilience of this animal." In Putnam County, New York, a similarly determined raccoon assaulted one victim at the end of her half-mile-long driveway. She held down the beast while attempting to free her cell phone to call the house; eventually, her husband and son had to club the raccoon repeatedly with a tire iron in order to kill it. ("I felt that nature had betrayed me," she later told a reporter for the public-radio show *This American Life*.) Then there was the red fox in South Carolina that pursued a nine-year-old as he walked to the school bus one morning. After an adult neighbor sheltered the boy, the fox latched on to the Good Samaritan's foot. He flung the animal into his home office, where it flailed against the walls and windows before finally falling asleep on a dog bed.

Nearly any species can be afflicted. Arizona officials were recently called to the scene after a dog was attacked by a mad peccary, a piglike creature whose residence in the Southwest had until that point been considered largely peaceable. In Robbins, North Carolina, it was a skunk that beset the pet Pekingese of David Sanders, who was forced to watch the two creatures battle it out for the better part of an hour. In Decatur County, Georgia, a donkey fell prey to the madness and bit its owner on the hand. In Imperial, Nebraska, the afflicted animal was literally a lamb, part of a child's 4-H project gone terribly, almost biblically awry. Some primeval force must truly be at work when the lamb can be made into a lion.

The agent of all these acts of possession is, of course, a virus. It is the most fatal virus in the world, a pathogen that kills nearly 100 percent of its hosts in most species, including humans. Fittingly, the rabies virus is shaped like a bullet: a cylindrical shell of glycoproteins and lipids that carries, in its rounded tip, a malevolent payload of helical RNA. On entering a living thing, it eschews the bloodstream, the default route of nearly all viruses but a path heavily guarded by immuno-protective sentries. Instead, like almost no other virus known to science, rabies sets its course through the nervous system, creeping upstream at one to two centimeters per day (on average) through the axoplasm, the transmission lines that conduct electrical impulses to and from the brain. Once inside the brain, the virus works slowly, diligently, fatally to warp the mind, suppressing the rational and stimulating the animal. Aggression rises to fever pitch; inhibitions melt away; salivation increases. The infected creature now has only days to live, and these he will likely spend on the attack, foaming at the mouth, chasing and lunging and biting in the throes of madness—because the demon that possesses him seeks more hosts.

If this sounds like a horror movie, we should not be surprised, for it is a scenario bound up into our very concept of horror. Rabies is a scourge as old as human civilization, and the terror of its manifestation is a fundamental human fear, because it challenges the boundary of humanity itself. That is, it troubles the line where man ends and animal begins—for the rabid bite is *the* visible symbol of the animal infecting the human, of an illness in a creature metamorphosing demonstrably into that same illness in a person.

Today, we understand that more than half our new diseases (60 percent, by a recent tally in *Nature*) are "zoonotic"—that is, originating in animal populations—and our widespread fear of the worst of these (swine flu, AIDS, West Nile, Ebola) has been colored by our knowledge of their bestial origins. It is hardly an exaggeration to say that nothing has made humans sicker than our association with animals. Not only our emerging diseases today but the major killers throughout the ages—smallpox, tuberculosis, malaria, influenza—evolved from similar diseases in animals. This is what Jared Diamond has called "the lethal gift of livestock," a major shaper of human destiny; the very fact that the farmer won out over the hunter-gatherer is due in part, Diamond argues, to the fact that the former "breathed out nastier germs." Through their close contact with animals, early farmers built up immunity to illnesses that would readily kill unexposed populations, a dynamic that still holds for emerging infectious diseases today.

Yet until the twentieth century, humans had no idea that so many of their illnesses derived from nonhuman hosts. During those years when the most catastrophic zoonosis in history struck—the fourteenth-century Black Death, or bubonic plague, which spreads to humans via fleas living on the backs of rats and other rodents—scholars blamed nearly everything else, from demonic forces and bad air to astronomical happenings and even human malefactors. For centuries, rabies was the only illness in which the animalistic transfer, or more like a transformation, was evident. No microscope was required to see the possession take place. A mad animal bit; a mad man appeared; each would die a terrible death. The madness could lurk within any mammal, even in—especially in—the most domesticated and loyal of all, the dog.

As the lone visible instance of animal-to-human infection, rabies has always shaded into something more supernatural: into bestial metamorphoses, into monstrous hybridities. When Greek myth beholds Lycaon, king of Arcadia, as he transforms into a slavering wolf, his countenance is "rabid," his jaws "besplutttered with foam." In fifteenth-century Spain, witch-hunters called *saludadores* were reputed, also, as healers of rabies, a convergence that made eminent sense given the widely held association between witches and their demonic canine "familiars." Between the fifteenth and the eighteenth centuries, Europe gestated two enduring legends whose part-human, part-animal villains bite their victims, thereby passing along their own degraded conditions—namely, the werewolf and the vampire, both of whom haunt the Western imagination to the present day. The essayist Susan Sontag noted that even as late as the nineteenth century, when viruses were becoming well understood and a rabies vaccine lay within reach, the true source of the rabies panics in France was not the fatality of the disease but rather the "fantasy"—though one might accurately say the *fact*—"that infection transformed people into maddened animals."

Paradoxically, during the twentieth century, after Pasteur's invention of a rabies vaccine provided a near-foolproof means of preventing its fatality in humans, our dark fascination with rabies seemed only to swell. The vaccine itself became as mythologized as the bug, such that even today many Americans believe that treatment requires some twenty (or is it thirty?) shots, delivered with a foot-long syringe into the stomach. (In fact, today's vaccine entails four shots, and not particularly deep in the arm.) Even as vaccination of dogs in the United States was reducing the infection rate in that species down to negligible levels, a generation of children learned to scrutinize their pet pooches for the slightest signs of madness, thanks in part to the lamentable influence of *Old Yeller,* a Walt Disney film about a frontier-era boy who falls in love with a yellow dog that becomes rabid. Twenty-four years later, a novel called *Cujo* (and its subsequent film adaptation) taught a whole new generation to fear rabies, albeit a bit more forthrightly: no one finished the book or left the theater surprised by what became of that nice dog.

It's almost as if the very anachronism of rabies, to the Western mind, has rendered it even more intriguing to us. Like the vampire, rabies carries with it the musty whiff of a centuries-old horror, even as it still terrifies us in the present day. Lately, TV comedians have taken to seizing on it for a laugh: two animated series created by Mike Judge, *King of the Hill* and *Beavis and Butt-Head,* have done episodes on rabies, as did the long-running medical comedy *Scrubs*. In the U.S. version of *The Office,* Michael Scott (the bumbling boss played by Steve Carell) tries to paper over the fact that he has hit an employee with his car by organizing a charity "race for the cure." The disease he chooses is rabies. He soon becomes perplexed, though, at how few donations are forthcoming:

> **Michael Scott:** I was also hoping to hand the giant check to a rabies doctor. How's that going?
>
> **Pam Beesly:** Not well. A doctor won't come out to collect a check for seven hundred dollars. Or five hundred dollars, if we go with the giant check. And also, there is no such thing as a rabies doctor.

Contrary to what your television may have told you, there are most assuredly still rabies doctors, and humans still die in the tens of thousands from the disease every year (fifty-five thousand, in the estimate of the World Health Organization). But few of these deaths happen in the United States or in western Europe. The dead hail overwhelmingly from Asia and Africa, from countries where vaccination is too expensive or too difficult to procure. And the course of their suffering is every bit as grim, and as inevitably fatal, as that endured by victims throughout the millennia.

Indeed, other than the wide availability of sedatives, which can subdue the final agonies of the disease, the sequence of horrors faced by a typical rabies patient today is hardly different from those experienced by the man who was probably the most eminent rabies victim in history: Charles Lennox, fourth Duke of Richmond, who for the two years leading up to his death in 1819 served as governor-general of Canada, the top post in what was then still a colonial government. The duke was a famous lover of dogs; a portrait of Lennox as a boy shows the young nobleman reclining against a tree stump as an adoring spaniel paws at his finery. Ironically, it was not a dog but rather a fox, the ostensibly tame pet of a soldier whose garrison the duke had occasion to inspect in Quebec, whose jaws were to blame for his demise. When the fox tangled with the duke's own dog—Blucher, so named in honor of Gebhard Leberecht von Blücher, the Prussian general who had recently bested Napoleon at Waterloo—Lennox manfully stepped in to separate the two. The mad fox seized this chance to insult the visiting dignitary, chomping down hard on the base of his thumb.

After a bite, the rabies virus binds quickly into the peripheral nerves but then makes its course with almost impossible sloth, usually requiring at least three weeks and often as long as three months to arrive at and penetrate the brain. On rare occasions a full year, or even five years, can elapse before the onset of symptoms. During this time the wound will heal over, and the victim may even forget about his scrape with a snarling beast. But healed or no, as the virus enters the brain, the wound will usually seem to return, as if by magic, with some odd sensation occurring at the site. This sensation can take many forms: stabbing pain, or numbness; burning, or unnatural cold;

tingling, or itching; or even a tremor. At roughly the same time, these soon-to-be-doomed patients typically display general signs of influenza, with a fever and perhaps a sore throat or some mild nausea. In the case of the Duke of Richmond, it began one day with shoulder pains and a sore throat, then progressed the following day to insomnia and fatigue.

All this is merely prelude to the illness itself, whose most notable symptom in humans—unique, as far as physicians know, to rabies among all diseases—is a terrifying condition called hydrophobia. As the term suggests, hydrophobia is a fear of water, though the word "fear" does not do justice to the eerie and fully physical manner in which it manifests. Present the hydrophobic patient with a cup of water and, desperately though he wants to drink it, his entire body rebels against the consummation of this act. The outstretched arm jerks away just as it is about to bring the cup to the parched lips. Other times the entire body convulses at the thought. Just beholding the water can make the diaphragm involuntarily contract, causing patients to gag and retch. On YouTube one can find video from a 2007 sufferer in Vietnam, showing the travails of a middle-aged construction worker as he attempts to consume some dark beverage from a clear plastic cup. He brings the vessel two-thirds of the distance to his lips before his hands begin to tremble uncontrollably. He stares at the fluid, mouth agape, his twitching hands sloshing it over the sides. Finally he forces himself to bring the tiniest sip into his mouth and, overcoming the revulsion in his gullet, to swallow it.

For the Duke of Richmond, though the chronology remains in some dispute, the hydrophobia seems to have struck first on the evening of August 26, 1819. At dinner with his officers, he found that his glass of claret disagreed with him. "I don't know how it is," he is said to have remarked to Colonel Francis Cockburn, one of his retinue, "but I cannot relish my wine tonight as usual. I feel that if I were a dog I should be shot for a mad one!"

The next day, the duke ate and drank almost nothing and remained in bed. By the evening, he found he could not drink at all. The following morning, a doctor prescribed a gargle, but this, too, had a "convulsive" effect on his throat. He could not even accept his customary shave, so repelled was he by the water in the basin. This day he dragged himself from his bed. He was scheduled to tour the swamps around the Ontario town of Richmond, recently renamed as such in his honor. But his body rebelled as he stepped into the boat. In terror he jumped back to the shore. Taken to the closest house, he begged to be moved farther inland: the very sound of running water had become unbearable to him. He was moved to a barn and laid down on a deathbed of straw.

Fevers spike high during this final phase of the disease. The mouth salivates profusely. Tears stream from the eyes. Goose bumps break out on the skin. Cries of agony, as expressed through a spasming throat, can produce the impression of an almost animal bark. In the throes of their convulsions, patients have even been known to bite. They also hallucinate. The eminent French physician Armand Trousseau, who practiced in the middle part of the nineteenth century, noted that "the patient is seized with sudden terror; he turns abruptly round, fancying that somebody calls to him." He cited the account of a colleague, one Dr. Bergeron, whose rabies patient "heard the ringing of bells, and saw mice run about on his bed."

Not uncommonly, male patients succumb to an even more lurid sort of abandon. The virus's action on the limbic system of the brain can cause them to exhibit hypersexual behavior: increased desire, involuntary erections, and even orgasms, sometimes occurring at a rate of once per hour. If the Duke of Richmond evidenced this symptom, his companions were too gallant to set it down for posterity. But other case reports from history describe up to thirty ejaculations in a single day. The Roman physician Galen, in his own remarks on rabies, describes the case of an unfortunate porter who suffered such emissions for three full days leading up to his death. Commenting on this grim fate the eighteenth-century Austrian physician Gerard van Swieten soberly noted, "*Semen et animam simul efflavit*": "His seed and his life were lost at the same time."

And yet, despite all the horrors of hydrophobia, arguably the most tragic aspect is the fact that the attacks will often subside, for a time, allowing sufferers periods of terrible, poignant lucidity: they are given the opportunity to fully contemplate what their condition portends. Before his death, the duke dictated a lengthy letter to his eldest daughter and also gave instructions that his beloved Blucher be handed over to her. "It will make her cry at first," he said, "but turn him in when she is alone and shut the door."

By now it should be apparent that this book is not for the squeamish or weak-kneed. Encounters with rabies have ever been thus. Louis Pasteur and his assistants, in order to develop their vaccine, had to corral dogs at the apex of their madness and extract deadly slaver from their snarling jaws. Axel Munthe, a Swedish physician, once saw Pasteur perform this trick with a glass tube held in his mouth, as two confederates with gloved hands pinned down a rabid bulldog. Some members of his team soon established a ghoulish fail-safe for these procedures. "At the beginning of each session a loaded revolver was placed within their reach," recalled Mary Cressac, the niece of Pasteur's collaborator Emile Roux. "If a terrible accident were to happen to one of them, the more courageous of the two others would put a bullet in his head."

We cannot claim so much bravado for this volume, on either our account or yours. A better analogy, perhaps, is the difficult process by which veterinarians submit suspect pets for rabies testing—another case study in how this diabolical disease causes nothing but agony for those who behold it. Even today, vets do not use a blood test for rabies in animals; it's not a pinprick and wait-and-see affair. Only a sampling from the brain will suffice. Therefore the animal must be killed, with its head removed and shipped off to authorities for study.

The first part of that process—capturing and humanely dispatching a deranged animal—is fairly standard stuff for your local vet. But carrying out a decapitation, even of a smallish creature, is much harder than they make it look in slasher pics. This is true not just for the obvious emotional reason: that in many cases the vet had been trying to save the life of this beloved pet just hours beforehand. It is also an ordeal in the purely practical sense. The cadaver is laid out on its back, contorted face canted skyward. With a scalpel the vet slices readily through the soft tissue around the animal's neck: fur and skin, muscles and vessels, esophagus and trachea.

Now the vet is stuck with the problem of the spine, the very conduit through which the rabies virus may—or may not—have passed; like Schrödinger's cat, the animal must be dead for this question to unravel. If the vet is lucky, her hospital has seen enough suspected rabies cases that it has thought to keep a hacksaw handy. In that case, she can take a brute-force path through bone, sawing straight through the tightly interlocked top vertebrae, the axis and the atlas. If she is not so lucky, she will have only her scalpel to work with. A five-minute job can thereby stretch out to twenty, as she is forced to disarticulate those two top backbones, severing the tendons that bind them and separating one from the other: a decidedly grisly brainteaser.

To be honest, our tour through the four-thousand-year history of rabies has felt a little like that. Sometimes whole weeks got lost in a blur of blood and fur. Our exploration into the cultural meaning of rabies took us deep into the gruesome medical case reports, from ancient and modern times. Then it flung us out again, into the murky realm of myth, to dog-headed men and zombie mobs and the mass butchering of Cairo's pigs. We've made pilgrimages to the Ardennes, to see the site of the holy rabies cure; to the rue d'Ulm in Paris, to behold the humble building where Pasteur performed his heroics; and to the island of Bali, where we finally came to stare the devil in the face ourselves.

Now, after two years of sawing, we feel we have finally finished the job, and we are pleased to ship it off to you, the reader. Come to think of it: in the case of a fox, or a cat, or even a toy-breed dog, the severed head might weigh just about the same as the book in your hands right now. Hold it in your outstretched palms, why don't you, and close your eyes. Not so very heavy, is it? And yet from packages this small—as suburban home owners sometimes learn, and as the Duke of Richmond discovered far too late—considerable mayhem can be unleashed.

Roman-era mosaic from England, depicting a wolf with Romulus and Remus.

1

IN THE BEGINNING

For more than a week, Achilles sulks while the Trojan War carries on without him. By just the third day of his absence, momentum has shifted decisively toward the Trojans, whose onslaught has repelled the invading Greeks back to their ships. As night falls, a Greek delegation, led by Odysseus, rushes to Achilles' encampment in hopes of luring him back into the scrum. The emissaries arrive to find the great hero listlessly strumming a lyre, warbling to no one in particular about the great deeds of warriors past. He is excited to have visitors. "Mix us stronger drink," he tells his henchman.

Libations poured, Odysseus lays out the Greeks' predicament. Their drubbing that day was accomplished almost entirely by one man, the Trojan hero Hector. Unlike Achilles, whose reputation in battle preceded him to the plain of Scamander, Hector has discovered his talent for killing more recently—"I have learned to be valiant," he remarks to his terrified wife. In this day's fighting he was particularly brilliant. When a Greek archer, aiming for Hector, killed instead his chariot driver, he leaped from the chariot, picked up a rock, and smashed the archer's collarbone, even as another arrow lay poised in his bow. Then Hector roused his army to drive the invaders back, his wild yet determined aspect in the chase resembling, in the words of the poet, "a hunting hound in the speed of his feet pursuing a wild boar or a lion."

In his pitch to Achilles, Odysseus describes Hector's battlefield ragings as "irresistible." And he attributes them, somewhat mysteriously, to a sort of *possession,* to a "strong fury" that has entered the Trojan hero. Hector has threatened to descend upon the Greek camp at dawn, to dismantle and burn their ships, and then, his quarry blinded by smoke, to dismantle the Greeks as well. Without Achilles, Odysseus warns, the Trojan might very well make good on the promise. If, however, Achilles returns, he will almost certainly slay Hector himself—for the very same "fury" has blinded Hector into believing that no man, not even Achilles, is his equal on the battlefield.

What is this peculiar fury that, in Odysseus's view, has possessed Hector, spurring him to unstoppable acts of martial courage but also to a mortal vulnerability? It is no ordinary anger. Homer's epics are awash in anger, with no fewer than nine terms employed to describe all the subtle flavors of fury. In *The Iliad,* this litany begins with the poem's very first word, *menin,* which so famously frames the entire epic around the "rage" of Achilles. But here in Odysseus's presentation to Achilles, the term for what has provoked Hector to such frenzy—*lyssa*—is something rather more primal. It has not been invoked anywhere in the poem before this scene, and with one notable exception the term will not appear again during the tale. It is a term closely linked to the word *lykos,* or "wolf," and is used to connote an animal state *beyond* anger: an insensate madness, a wolfish rage. Later, in tragedies, Lyssa is sometimes personified, goading Heracles to slay his family and Pentheus's own mother and aunt to dismember him. Vase paintings occasionally depict her as a feminine form wearing a dog's head as a cap.

In the realm of epic and myth, *lyssa* is impossible to properly define. In the factual prose of Attic Greece, however, the word had a quite literal meaning: rabies. As much as we hesitate (obeying the injunction of Susan Sontag) to deploy illnesses as metaphors, such links can hardly be resisted even in the present day, when the emergence of new diseases—usually originating in animal populations—threatens us with unforeseen manners of death. Consider how inconceivable it would have been to disentangle such links at a time before men knew of viruses, a time when diseases spread by means the keenest eye could not discern nor the keenest mind divine. With this particular convergence, the twinning of rabies with notions of savage possession, it is hard even to say which member of the pair took precedence, chronologically or otherwise. Both were there from the beginning. To link the two states, medical and metaphorical, was natural in both senses of that word. *Lyssa* was rare, terrifying; violent, and animalistically destructive of others; ultimately (and pathetically) destructive of self. It made creatures maim and kill those closest to them. It hollowed out reason and left nothing but frenzy.

After Odysseus's speech to Achilles, *lyssa* makes one last, dramatic entrance in *The Iliad*. Though he resists the entreaty at first, Achilles does eventually return to the fight. He leads the Greeks to the gates of Troy, which open to shelter the Trojan warriors in their desperate retreat. Odysseus's prediction is destined to come true: Hector, unmoved by the pleas of his parents, will wait outside the city gates the next day—"as a snake waits for a man by his hole, in the mountains, glutted with evil poisons"—intent on doing combat against the Greek hero alone. On the eve of this fateful encounter, as the king of Troy opens the gates of his doomed city, Achilles pursues the fleeing Trojans with spear aloft, and the "powerful *lyssa* unrelentingly possesses his heart."

Rabies has always been with us. For as long as there has been writing, we have written about it. For as long, even, as we have kept company with dogs, this menace inside them has sometimes emerged to show its face to us. But perhaps the most impressive sign of its longevity is this: rabies serves as the setup for one of humanity's first recorded jokes. (Stop us if you've heard it before.)

A Babylonian fellow gets bitten by a dog. He travels to Isin, renowned city of the goddess of health. There, a high priest recites an incantation upon him, and the patient is very pleased with the quality of care.

"May you be blessed for the healing you have done!" the visitor cries. "You must come to Nippur, where I live. I'll bring you a coat, carve off the choicest cuts for you, and give you barley beer to drink—two jugfuls!"

Perhaps to his surprise, the priest takes him up on the offer. "Where in Nippur shall I come?" he asks.

"Well," the patient continues, his voice betraying some hesitation, "enter by Grand Gate . . . keep Broad Avenue, the boulevard, and Right Street on your left. A woman named Beltiya-sharrat-Apsî, who tends a garden there, will be sitting at a plot selling vegetables. Ask her and she will show you."

Despite these suspiciously difficult directions (keep Right Street on the left!), the old doctor somehow arrives at the garden in question. But it turns out that Beltiya-sharrat-Apsî is the most unhelpful woman alive. And because she speaks such a thick Nippurian dialect, she and the holy man from Isin can hardly communicate. After a tortured exchange, the old woman gives up on the priest in exasperation, and the patient (we are left to presume) is never forced to make good on his promised co-pay.

No doubt this gag, found inscribed on a clay tablet, was funnier in the original Akkadian. But baked into its premise is an obvious question: Why, over an injury as straightforward as a dog bite, would a patient venture all the way from Nippur to Isin, a distance of nearly twenty miles, in order to see a healer? And not just any healer: the cuneiform text indicates the high priest to be none other than the main administrator (*šangû*) of the foremost temple of Isin—which, again, was the city of the goddess of medicine. Having been bitten by a dog, this Babylonian has rushed off to the equivalent of the Mayo Clinic.

This could very well be part of the joke. Yet ample evidence exists in the records of early Mesopotamia that dog bites were feared for a very rational and very terrifying reason. Nearly two thousand years before Christ, the Laws of Eshnunna—a precursor to the Code of Hammurabi—stipulate punishment for the owner of a *kalbum šegûm,* or "rabid dog": "If a dog becomes rabid and the ward authority makes that known to its owner, but he does not watch over his dog so that it bites a man and causes his death, the owner of the dog shall pay forty shekels of silver; if it bites a slave and causes his death, he shall pay fifteen shekels of silver."

Contemporary Assyriologists have found references to rabies in private letters ("Like a rabid dog, he does not know where he will bite next"); in the omens of entrails readers (a hole in a particular section of the animal's liver indicated that a man would contract rabies); in astrology (lunar eclipses in particular months were said to portend outbreaks among dogs); and in the Marduk Prophecy, an apocalyptic text from the first millennium B.C. in which Marduk, then the preeminent deity, threatens to abandon Babylon and thereby unleash a series of plagues, the last of these being rabies:

> I will send the gods of cattle and grain off to the heavens. The god of beer will make ill the heart of the land. The corpses of people will clog the gates. Brother will eat brother. Friend will kill friend with a weapon. . . . Lions will cut off the roads. Dogs will become rabid and bite people. All the persons whom they bite will not survive but will die.

Rabies also figures in some of the incantations that were used, as by the high priest of Isin, in attempts to cure disease. One Babylonian list of maladies, in grouping the dog bite together with the scorpion sting and the snakebite, describes the canine affliction as "the bite that grows up." The surviving incantations against this bite use a curious metaphor to describe what the dog's jaws have left in the wound: the dog's "semen is carried in his mouth," and "where it has bitten, it has left its child." (Another incantation expresses this with great pith: "May the bite of the dog not produce puppies!") Given how many contingencies exist in the transmission of the rabies virus—whether the animal is actually rabid and not merely vicious, whether its bite has actually punctured the skin, whether the virus takes root in the nervous system and begins its climb to the brain—Mesopotamian doctors would often have had reason to believe their spells had cured the patient. "Remove the madness from his face and fear from his lips!" one spell exhorted. "Let the dog die and the man survive!" cried another.

One Sumerian incantation against rabies tells the priest to work magic on purified water, which the patient is then compelled to drink: "Cast the spell into the water! Feed the water to the patient, so that the venom itself can go out!" Given that hydrophobia has presented throughout history as the defining symptom of the disease, this prescribed treatment is a bit ironic.

Western histories of medicine tend to favor the Greeks—and understandably so, given the legacy of Hippocrates (ca. 400 B.C.) and the generations of medical authors who expanded on his wisdom in the centuries that followed. But arguably the most impressive description of rabies from the ancient world appears in the *Suśruta samhita,* a classic text of Ayurveda, the Indian system of traditional medicine. By and large, authorship of this tome is attributed to its namesake, Suśruta, who practiced medicine in the city of Varanasi, on the banks of the Ganges. Most contemporary historians place Suśruta in the first or second century A.D., though some put him far earlier, and that is merely the beginning of our complications in affixing a date to the work. On the one hand, the text was edited, apparently heavily, by a later disciple named Nāgārjuna, who lived sometime between the fifth and the tenth centuries A.D. On the other hand, the *Samhita* is said to collect wisdom from Suśruta's hallowed ancestor, Divodāsa Dhanvantari, who by all accounts lived in 1000 B.C. or earlier.

Regardless, the *Samhita* is a stunning medical compendium for its time. Suśruta is best known to medical historians as the father of surgery, and indeed the *Samhita* documents a staggering array of procedures, from the nose job to the cesarean section. For the eye alone, Suśruta's text includes eighteen chapters laying out some fifty-one different operations, many of them quite sophisticated in the details.

The *Samhita* devotes nearly a thousand words to rabies, and based on the versions of the text that have been handed down, it correctly identifies many aspects of the disease. It recognizes that humans contract the disease and can be said to display comparable symptoms: "A person bitten by a rabid animal barks and howls like the animal by which he is bitten," causing him to lose the "functions and faculties of a human subject." The ancient Ayurvedans also claim the honor of being the first early medical scholars to isolate the phenomenon of hydrophobia itself and to recognize that a human illness progressing to that phase is invariably fatal: "If the patient in such a case becomes exceedingly frightened at the sight or mention of the very name of water, he should be understood to have been afflicted with Jala-trása [water-scare] and be deemed to have been doomed." (The *Samhita* goes a bit too far, however, by claiming that Jala-trása is fatal even in patients who are merely frightened by *reports* of Jala-trása.)

Although the disease is presented as an interaction of the "wind" of the human and the "phlegm" of the animal—Vāyu and Kapha, respectively, concepts that live on today in the practice of Ayurveda—the *Samhita* recognizes that rabies is fundamentally neurological in nature, and it also provides a fairly reliable description of rabies as it manifests in animals:

> The bodily Vāyu, in conjunction with the Kapha of a jackal, dog, wolf, bear, tiger or of any other such ferocious beast, affects the sensory nerves of these animals and overwhelms their instinct and consciousness. The tails, jaw-bones, and shoulders of such infuriated animals naturally droop down, attended with a copious flow of saliva from their mouths. The beasts in such a state

of frenzy, blinded and deafened by rage, roam about and bite each other.

Among the vaunted Greeks, the medical understanding of *lyssa* was not nearly so sophisticated. Reference to the disease does not appear explicitly in Hippocrates. Aristotle does address rabies directly in his treatise *History of Animals,* though he flubs it in nearly every respect. Dogs, he wrote with an odd confidence, suffer from only three diseases: *lyssa,* or rabies; *cynanche,* severe sore throat or tonsillitis; and *podagra,* or gout.* The philosopher also held the belief that rabies could not be contracted by humans: "Rabies drives the animal mad, and any animal whatever, excepting man, will take the disease if bitten by a dog so afflicted; the disease is fatal to the dog itself, and to any animal it may bite, man excepted." (Aristotle added that the elephant, generally thought then to be immune to disease, "is occasionally subject to flatulency.")

In the first two centuries A.D., during roughly the same period that Suśruta (if the consensus estimates are true) was practicing his surgical wonders, the Greco-Roman tradition of medicine did begin to develop a comparably sophisticated understanding of rabies. This awareness begins with Aulus Cornelius Celsus, believed to have been born in around 25 B.C. and to have written during the early part of the first century. Almost nothing is known of Celsus's life, in a civilization that took great pains to document the lives of men it considered sufficiently worthy. Pliny the Elder, the first-century historian and naturalist, believed that Celsus lived in the southern part of France, based on his reference to a vine that was native to that region. Celsus seems to

---

* In the philosopher's defense, R. H. A. Merlen, author of a fine volume entitled *De Canibus: Dog and Hound in Antiquity,* surmises that *cynanche* was actually itself a form of rabies—so-called dumb rabies, in which the afflicted dog, rather than raging, stands mute with its mouth agape. Merlen points out that Aristotle characterizes *cynanche* as fatal in dogs, unlike any commonly presenting throat malady.

have been not a physician but an encyclopedist who compiled his *De medicina* largely from the Greeks. But while his notes on hydrophobia go well beyond the silence of Hippocrates and the complete misapprehension of Aristotle, they are hardly more incisive. He does recognize the existence of hydrophobia ("a most distressing disease, in which the patient is tortured simultaneously by thirst and by dread of water") and pays at least lip service to the fact that "in these cases there is very little help for the sufferer." His description of the malady ends there, though, and the balance of his account is given over to an elaborate and darkly amusing series of treatments—more on which later.

It was a hundred years or so after Celsus that a school of scientific thought emerged to spur the classical tradition toward a better understanding not just of rabies but of medicine as a whole. Hoping to escape the intellectual strictures of the empiricists—who rejected not only experimentation but all theoretical approaches to medicine, holding that physicians should work based only on what they could perceive with the naked eye—these scholars called themselves the methodists, and they put forward a positive theory of how the human body functioned. That their theory (which involved conceiving of diseases as "affections" and considering their effects holistically) strikes the contemporary mind as largely nonsensical seems to have been beside the point. The methodists' focus on improving therapy invigorated the whole enterprise of writing and thinking about human health.

The founder of the methodist school, Themison (first century B.C.), and one of his disciples, Eudemus, were both said to have survived attacks by rabid dogs; and either might have been the original author of an anonymous methodist text, usually dated to the first century A.D., that touches on rabies at length. More impressive still are the notes on hydrophobia made by Soranus, a methodist physician (first or second century A.D.) from Ephesus, on the western shore of what is now Turkey. Best known today for his prescient thoughts on gynecology, Soranus also left behind—in his treatises on acute and chronic diseases, which survive in a full Latin translation made by Caelius Aurelianus during the fifth century—a few fairly lengthy sections on hydrophobia and its treatment. Unlike most of his predecessors, Soranus recognized that contact with rabid animals is the only means by which hydrophobia spreads. He even gives a seemingly far-fetched example that in fact might be possible, given what we know today about the disease:

> And once when a seamstress was preparing to patch a cloak rent by the bites of a rabid animal, she adjusted the threads along the end, using her tongue, and then as she sewed she licked the edges that were being joined, in order to make the passage of the needle easier. It is reported that two days later she was stricken by rabies.

One gathers that Soranus is drawing upon close observation of quite a few rabies sufferers, given his thorough and plausible list of symptoms, which includes not just revulsion at water but rapid and irregular pulse, fever, incontinence, shaking, and—noted for the first time—involuntary ejaculation. He correctly rebukes an earlier writer for having claimed the disease can sometimes progress over the course of years. He even rebuffs Eudemus, Themison's successor, for his idea that melancholy and hydrophobia were one and the same. "The victims of hydrophobia die quickly," Soranus writes, "for it is not only an acute disease but one that is unremitting."

Throughout this vast expanse of history, the constant threat of rabies—rare yet tinged with horror—served as merely another wrinkle in early civilization's intimate, complex relationship with the dog. By this point dogs had been domesticated (or had domesticated themselves, as most scholars now believe) for at least ten thousand years, and yet their role in society was profoundly dissonant. Based on findings of teeth and bones around Mesopotamian sites, archaeologists have concluded that semi-feral dogs roamed cities as scavengers, feeding on trash. Yet many dogs were companions, for whose actions a human was responsible, as the Laws of Eshnunna attest. Alongside the other attendant advances of civilization, such as the city and the written word, reliable breeding of dogs first emerged during this period: remains of sight hounds—a purebred line of dogs that persists to this day in such graceful runners as greyhounds, whippets, and salukis—have been found in the region dating back as far as 3500 B.C. Dogs also figured prominently in the spiritual symbology of early Mesopotamia. In a curious connection to our joke, the dog seems to have been most invoked as the symbol of Gula, the spirit of healing; when archaeologists excavated her temple at Isin, the very same one over which the high priest is said to have presided, they found it studded with dog figurines. Later, King Nebuchadnezzar II, writing in roughly 600 B.C., records that dog statuettes made of precious metals were left at Gula's temple in Babylon.

And so dogs occupied the lowest and yet also the highest rungs in the bestiary of early man: pitiful scavenger of garbage but also huntmate, totem, friend. It is a dual nature that persists into the present day, as one can vividly witness on the streets of any developing-world city, where the collared and the bathed uneasily coexist with the unkept and the unkempt. It is also a bifurcation that dates all the way back to the beginning of dog history, to the very first creatures to earn the name *Canis familiaris.* Scientists theorize that the indispensable hearth of domestication was the human garbage pile, with the wolves that scavenged there some fifteen thousand years ago becoming gradually more tame. By studying the mitochondrial DNA of various dogs from around the world, geneticists have tracked the site of this domestication to southern China—which means that, based on local traditions and archaeological records, the first dogs may have been bred for use as food. Dog bones found by archaeologists in that region are often scarred with knife marks.

Nevertheless, the dog almost immediately became something

more. The uncanny ability of dogs to pick up on human moods and needs is almost certainly instinctive rather than bred, so we can imagine the slow courtship of human and beast as it would have played out over centuries. Without even having to be captured or trained, some tamed dogs would have begun to function as guard animals, alerting humans to potential intrusion, protecting food and other possessions from outside assault. Soon, with training, these ur-pets would have been hunting, pulling sleds, and, eventually, herding livestock—man and dog, creating civilization as one.

Yet the hand that feeds the dog has forever been not merely bitten by it but, on occasion, devoured. In rabies, after all, every dog has a dark side lurking behind the soulful eyes; and even a healthy dog seldom hesitates to feast on the corpse of a dead human, even that of a former friend or master. In ancient India, the ambivalence toward dogs was eloquently expressed in one sacred text called the *NisīhaCū,* which says that gods "come to the world of men in the shape of *yaksas,* dogs, that is. They are worshipped when they do good, and not, when they do not." Indian literature itself is rife with images of dogs as battlefield scavengers; in one sacred text, hell is portrayed as a place where malign rulers are devoured by 720 dogs with fangs of steel. And yet dogs were kept as pets and bred by the elite, and the favors of a dog were sometimes auspicious. "If a dog comes face-to-face with [a man] in a joyous mood," noted one ancient work, "frolicking and rolling on the ground in front of him, then . . . there will be a great gain of wealth [when he] starts on a journey."

Nowhere was the dichotomy in the dog starker than among the ancient Egyptians, whose highest god was the dog god Anubis and who bred graceful sight hounds—the lithe form of which is believed by some to survive to the present day in our pharaoh hounds. An excavated tomb at Abydos, dating to 3300 B.C., built during the pre-pharaonic Upper Kingdom for an unknown ruler, shows evidence of the ritual burial of dogs, which would become a common practice in Egypt. A tomb at Hierakonpolis from approximately the same time (discovered during the late nineteenth century but then lost) was illustrated in full color with a hunting scene, complete with hounds, and contained the remains of multiple domesticated dogs; excavation of the tomb of Queen Herneith, who ruled a few hundred years later, found the skeleton of her dog stretched across the entrance to her tomb, guarding her home in the afterlife. In art, hounds were often depicted on leashes and as widely present in human society; ancient Egypt was a dog's paradise, a place where (if we are to believe Herodotus) the death of a pet dog would prompt its owner to shave not just his head but his entire body.

And yet even in Egypt, semi-feral dogs posed a constant threat in the streets of towns and villages; in *The Book of the Dead,* dogs are alluded to in one appeal by the deceased narrator to Ra, the sun deity, against a force that "carries off souls, who gulps down decayed matter, who lives on carrion, who is attached to darkness and dwells in gloom, of whom the feeble are afraid." References to dogs as scavengers in Egypt are found not just in the Hebrew Bible's accounts but in papyri documenting the Roman era there.

Like the Egyptians, the Greeks loved their graceful hunting hounds and considered them loyal friends and companions. A new literary genre, the *cynegeticon,* sprang up in ancient Greece to extol the hound and to prescribe its proper breeding and care. The most prominent (and likely first) of these guidebooks was penned by the soldier-historian Xenophon, who himself had witnessed the power of *lyssa* during a military campaign: of a fleeing enemy he wrote, with a hint of boast, "They were afraid that some *lyssa,* like that of dogs, had seized our men." After he was exiled from Athens to the Peloponnesian town of Scillus, Xenophon spent his postmilitary years in a happy reverie of hunting and writing, pursuits that converged in his *Cynegeticus.* He describes his ideal hounds in sumptuous detail: flat and muscular head, small thin ears, long straight tail, sparkling black eyes; the forelegs "short, straight, round and firm"; the hips "round and fleshy at the back, not close at the top, and smooth on the inside"; the hind legs "much longer than the forelegs and slightly bent." Profound respect suffuses every line of the *Cynegeticus*. Xenophon chastens the hunter not to employ collars that might chafe the dog's coat. He prescribes the praise that should be showered upon the hounds while they chase the hare. "Now, hounds, now!" one is enjoined to shout. "Well done! Bravo, hounds! Well done, hounds!"*

Nevertheless, scavenging dogs also roamed Greek fields and towns, carrying upon them the stench of death. *The Iliad* invokes the dog perhaps twenty times as a devourer of corpse flesh, the first instance occurring in the second sentence of the epic's very first stanza: "Many a brave soul did [the anger of Achilles] send hurrying down to Hades, and many a hero did it yield as prey to dogs and vultures." Hector's father, the old king Priam, captures the sad irony of the fate that lies in store for him as he contemplates his imminent death at the hand of Achilles. "My dogs in front of my doorway," he foretells,

> *will rip me raw, after some man with stroke of the sharp bronze*
> *spear, or with spearcast, has torn the life out of my body;*
> *those dogs I raised in my halls to be at my table, to guard my*
> *gates, who will lap my blood in the savagery of their anger*
> *and then lie down in my courts. For a young man all is decorous*
> *when he is cut down in battle and torn with the sharp bronze,*
> *and lies there*
> *dead, and though dead still all that shows about him is beautiful;*
> *but when an old man is dead and down, and the dogs mutilate*

* Xenophon even enumerates, at humorous length, a list of ideal names for hounds: Psyche, Pluck, Buckler, Spigot, Lance, Lurcher, Watch, Keeper, Brigade, Fencer, Butcher, Blazer, Prowess, Craftsman, Forester, Counsellor, Spoiler, Hurry, Fury, Growler, Riot, Bloomer, Rome, Blossom, Hebe, Hilary, Jolity, Gazer, Eyebright, Much, Force, Trooper, Bustle, Bubbler, Rockdove, Stubborn, Yelp, Killer, Pêle-mêle, Strongboy, Sky, Sunbeam, Bodkin, Wistful, Gnome, Tracks, Dash—"short names," he reasons, "which will be easy to call out."

> *the grey head and the grey beard and the parts that are secret,*
> *this, for all sad mortality, is the sight most pitiful.*

The word "dog" was also hurled as an epithet to decry the shameless man or woman; *The Iliad* finds Iris slinging it at Athena and Helen of Troy applying it ruefully to herself.

Beyond the dog's fondness for corpse flesh, it also could succumb at any time (literally or metaphorically) to the frenzied madness of *lyssa*. One need look no further than the mythic fate of Actaeon, the hunter whose severe misfortune it is to stumble across Diana, goddess of the hunt, as she bathes in the woods. To punish him, she turns him into a stag, prompting a second, bestial transformation that causes his death: his own beloved hounds, seized by *lyssa* at the sight of his new form, set upon him and tear him limb from limb.

Ovid, in the *Metamorphoses*—an all-encompassing volume about human-to-animal transformations—renders both transitions with awful acuity, allowing us to experience both from inside the hunter's still-human consciousness. Actaeon realizes he has become a stag only when he witnesses his reflection in a pool. "Poor me!" he tries to exclaim at the sight but manages only to emit a groan, and thereby learns that groaning, for him, "was now speech." His body has become alien to him—"tears streamed down cheeks that were no longer his"—even as his mind is left untouched, permitting him to grasp the full horror of his situation.

Almost immediately thereafter come the hounds, formerly his charges but now his pursuers, "rushing at him like a storm." His conscious mind lingers on each of them, one by one, noting their names and, at times, an endearing bit of detail that only an owner could know: Speedy and Wolf are siblings, while Shepherdess leads two puppies from a recent litter; Sylvia has "lately been gored by a boar." Some thirty-five dogs are noted by name, with "many more too numerous to mention," all dogs he has raised and fed; now they charge toward him in a slavering mob, "out to taste his blood." It is hard to know which of these twinned faces of *lyssa* is more horrible, in either Ovid's reckoning or ours: the human becoming animal, or the hunter being hunted by his own treasured dogs.

Perhaps the most enduring ancient symbol of the dog's two warring natures is Cerberus, that terrifying watchdog whose vigilant gaze and fearsome jaws kept the dead from escaping Hades and returning to the world of the living. Descriptions of his physiology vary significantly in the different retellings—his heads number two, sometimes three, sometimes fifty, or even a hundred; his tail is that of a snake, or not; snake heads sometimes sprout from his head and neck like a gruesome mane. But despite all these monstrous innovations he is consistently described as a dog. A "cursed" or "dreaded" or "savage" dog he may be, but he remains a dog nonetheless, the unmistakable kin of those that walk the earth and lick its inhabitants. He could even be a good dog, at times. As described by Hesiod, Cerberus was quite friendly to the dying, at least when they arrived; he positively welcomed them, in fact, "with actions of his tail and both ears." It was only when they attempted to pass *back* into life that he would set upon them savagely, even devour them. Death is a boundary that can be freely crossed in only one direction, and so guarding that boundary is a perfect role for a dog: natural friend on the one hand—or head; savage attacker and corpse devourer on the other; both natures cohabiting inside one vexing four-footed form.

It was more than just the power of Cerberus's many jaws that was to be feared. In the *Metamorphoses,* a list of poisonous substances includes "slaver from Cerberus," along with a creation myth whereby that rabid saliva, sprayed from the hellhound's lips and flecking a field of battle, gave rise to a notoriously poisonous plant called aconite—also known, tellingly, as wolfsbane. As the veterinary historian John Blaisdell has noted, symptoms of aconite poisoning in humans bear some passing similarity to those of rabies: they can include frothy saliva, impaired vision, vertigo, and finally a coma. It is not improbable that some ancient Greeks would have believed that this poison, mythically born of Cerberus's lips, was literally the same as that to be found inside the mouth of a rabid dog.

Until just the past century—and even then only in the developed world—rabies has been experienced by humans as a disease of the dog, a peculiarly canine madness that could reproduce a similar, fatal madness in humans. But all the while, the disease also lurked inside another, far more shadowy species: the bat. Indeed, recent research has indicated that bats harbored the disease even earlier than dogs, going back at least seven thousand years and as far as twelve thousand years, far before the first written languages and perhaps even before dogs were domesticated from wolves.

How was this calculation made? The answer flows from two simple facts about how viruses evolve over time. The first is that most mutations in a virus are neither beneficial nor harmful to its propagation; instead, they're neutral, trivially altering the genetic sequence without changing the virus's overall fitness in any way. The second fact is that these mutations tend, over large populations and long periods of time, to happen on a predictable schedule. So given a set of related viral strains, a computer can analyze the patterns of genetic difference and arrange them into a rough phylogenetic tree, showing which strain evolved from which and how long ago the divergences occurred. In 2001, two researchers at France's Institut Pasteur used this technique to investigate a large set of rabies virus strains—thirty-six from dogs and seventeen from bats—and the results were fairly clear: the enigmatic bat, a distant presence for most of the cultural history of rabies, was probably responsible for infecting the dog, rather than the other way around.

This so-called molecular clock research has led to many other insights about the origins of disease. In particular, it's shown us how many of our worst killers, pathogens that have racked humanity since the earliest civilizations, evolved out of animal populations. Measles, we now know, evolved from a disease in cattle; similarly, the various strains of influenza, as we still see today in our annual flu scares, readily pass back and forth between us and our livestock (for more on this, see Chapter 6). Some of these zoonotic leaps from animal to man have been understood fully only during the past decade or so, as genome sequencing has allowed scientists to trace more precisely the genetic lineage of pathogens. For example, a team led by the Stanford epidemiologist Nathan Wolfe announced in 2009 that it had isolated the origins of malaria in a parasite of chimpanzees, which presumably spread to humans through mosquito bites.

New sleuthing has yielded particularly intriguing details about smallpox, arguably the deadliest disease in history. A 2007 study, headed up by researchers at the U.S. Centers for Disease Control and Prevention, traced the notorious killer back to a virus in rodents, estimating that it made the leap to humans at least sixteen thousand years ago. What is especially satisfying is the team's identification of two separate human strains, an earlier and milder version that cropped up in west Africa and the Americas, and a more severe version—the progenitor of the strain that slew untold millions over the past millennium before its eradication in the late 1970s—that emerged from Asia a bit later. This helps explain why the literature, medical and otherwise, of the Greeks and Romans provides little evidence that highly fatal smallpox was common, even though archaeological evidence shows the clear presence of a smallpox-like condition in ancient Egypt. The most spectacular example of this is the mummified body of Pharaoh Ramses V, on whose shriveled skin can clearly be seen the pustular pattern typical of the disease. (This possibly answers the vexing Egyptological question of why Ramses V was not buried for almost two years after his death, when other pharaohs were interred just seventy days after mummification; either fear of infection from his corpse or a paucity of healthy embalmers might account for the lag.)

Smallpox was far from the only ancient epidemic with its origins in the rodent. Both plague and typhus ravaged by way of the rat, whose fleas would transmit the deadly bugs to unsuspecting humans. For all the emphasis placed on livestock in the development of civilization, the case can be made—and indeed has been made, most elegantly by the biologist Hans Zinsser in his 1935 book *Rats, Lice, and History*—that human affairs have been stirred far more vigorously by the rat, whose companionship with people has tended to be involuntary on our part but whose omnipresence among us, like that of the stray dog, became more or less inevitable with the emergence of the city. With most zoonotic leaps in disease, animal contact is the spark, but urbanization is the bone-dry tinder; a newly evolved pathogen can't spread from person to person, after all, unless people run across one another in the first place.

How to treat the rabies patient or the dog-bite victim? Consider the predicament of an ancient physician on this terrible question. The cause of hydrophobia (the bite of a rabid animal) was often separated by many weeks from its effect (the onset of neurological symptoms), and only a fraction of bites—even assuming an animal that is actually rabid and not merely vicious—progressed to the fatal infection. Meanwhile, it was hard to distinguish real cases of hydrophobia from hysterical ones, which were common right up to the twentieth century. Worse, because of the relative paucity of cases, ancient medical scholars often compiled alleged cures from second- and thirdhand reports.

For all of these reasons we should forgive, at least to a point, the extraordinary nonsense that passed for rabies treatment in the ancient world. Let's begin with bite treatment. Here again the *Suśruta samhita* deserves the most respect. Not only does it acknowledge, without wavering, the fatality of hydrophobia, but it prescribes a treatment for rabid bites—bleeding and cauterization of the wound—that is as sensible as any. (Also as delicious as any: the *Samhita* recommends cauterizing with clarified butter, which the patient is then invited to drink. It also prescribes a sesame paste for the wound and advises that the patient be fed a special fire-baked cake made of rice, roots, and leaves. The Varanasian patient did not face death on an empty stomach.)

In ancient China, where mentions of rabies in extant texts are relatively spare, the disease does appear in Ge Hong's "Handy Therapies for Emergencies," from the third century A.D. Ge prescribes "moxibustion" for the wound, a process that involved burning mugwort, a species of wormwood, and applying it to the bitten region. This was likely to have been more effective, or at least to do less harm, than another of his recommendations: to kill the offending dog, remove its brain, and rub that on the wound.

Among the Greco-Romans, perhaps we should not be surprised that Celsus, the encyclopedist, drawing as he did on many different sources, some of uncertain provenance, should supply us with a far more varied list of dog-bite treatments. These include bleeding and cauterization, but also the application of salt, or even a brine pickle, to the wound. Some physicians, he says, send their patients to a steam bath, "there to sweat as much as their bodily strength allows, the wound being kept open in order that the poison may drop out freely from it." After that, the doctors pour wine into the bite. "When this has been carried out for three days," Celsus says, "the patient is deemed to be out of danger."

Things totter off the rails with Pliny the Elder. As with Ge Hong, Pliny's thoughts tend to involve using the animal to treat the man. His best-known cure—to "insert in the wound ashes of hairs from the tail of the dog that inflicted the bite"—lives on today in our expression "hair of the dog," referring to a not-quite-so-dubious hangover remedy. But Pliny thought that a maggot from any dead dog's carcass would do the trick, as would a linen cloth soaked with the menstrual blood of a female dog. Or the rabid dog's head could be burned to ashes, and the ashes applied to the wound; or the head could just be eaten outright.

Still not see a treatment that works for you? Let Dr. Pliny lay out some more options:

> There is a small worm in a dog's tongue . . . : if this is removed from the animal while a pup, it will never become mad or lose its appetite. This worm, after being carried thrice round a fire, is given to persons who have been bitten by a mad dog, to prevent them from becoming mad. This madness, too, is prevented by eating a cock's brains; but the virtue of these brains lasts for one year only, and no more. They say, too, that a cock's comb, pounded, is highly efficacious as an application to the wound; as also, goose-grease, mixed with honey. The flesh also of a mad dog is sometimes salted, and taken with the food, as a remedy for this disease. In addition to this, young puppies of the same sex as the dog that has inflicted the injury, are drowned in water, and the person who has been bitten eats their liver raw. The dung of poultry, provided it is of a red colour, is very useful, applied with vinegar; the ashes, too, of the tail of a shrew-mouse, if the animal has survived and been set at liberty; a clod from a swallow's nest, applied with vinegar; the young of a swallow, reduced to ashes; or the skin or old slough of a serpent that has been cast in spring, beaten up with a male crab in wine.

"This slough," Pliny adds, "put away by itself in chests and drawers, destroys moths."

To the credit of Pliny, and of Celsus (with one exception, below), all these proffered treatments address the rabid dog or its bite, not hydrophobia itself. But even the fatal manifestation of the disease occasioned some elaborate and entirely chimerical cures. Oddly, the methodists, whose observations about hydrophobic symptoms became increasingly admirable over the centuries, seem to get more addled when the subject is treatment. Both the anonymous Greek text and, later, Soranus himself wrote of treatment as if recovery were more likely than not. They recommended creating a spa-like atmosphere. "Have patients suffering from hydrophobia lie in rooms with good air well tempered," remarked the anonymous author. "Massage his limbs," added Soranus, and "cover with warm, clean wool or cloths those parts that are affected by spasm." Both authors presented hydrophobia as an acute attack that would often recede in time—a bewildering judgment that flies in the face of observable facts. They prescribed various poultices made from dates crushed with quinces, or olive oil, or ripe melon, or vine tendrils, or coriander. Some unnamed physicians, cited by Soranus, recommend that a plaster be made from hellebores—the flowering perennials—and applied to the anus.

The most remarkable, and perhaps fitting, prescription for hydrophobia is the one offered by Celsus, who, as noted previously, had the good sense to admit that there was "little help" for the hydrophobic patient at all. Yet he apparently could not refrain from offering just one little cure: that is, "to throw the patient unawares into a water tank which he has not seen beforehand." He explains this method to be, as we might say today, win-win:

> If he cannot swim, let him sink under and drink, then lift him out; if he can swim, push him under at intervals so that he drinks his fill of water even against his will; for so his thirst and dread of water are removed at the same time.

If this proto-waterboarding happens to spur muscle spasms in the subject, Celsus recommends he be "taken straight from the tank and plunged into a bath of hot oil." A patient could be forgiven for preferring hydrophobia to that particular fate.

The basilica at Saint-Hubert, 2010.

2

THE MIDDLE RAGES

On the subject of Saint Hubert, protector of hunters, healer of rabies sufferers, we might as well begin with the myth; for while the truth about his life remains stubbornly opaque, it was the myth, and not the truth, that brought generations of fearful dog-bitten pilgrims from across Europe, by foot and horse and eventually even train, to be cured at the site where his holy relics resided.

The myth begins when Hubert is a young seventh-century nobleman in the Frankish kingdom, son of the Duke of Aquitaine. Hubert decides to spurn courtly life. He retreats to the deep forests of the Ardennes, the range of noble hills that roll through what is now Belgium and into the east of France, and devotes himself to the hunt. One Good Friday, so the legend goes, the young man is giving chase to a stag when the beast rears around, a crucifix hovering between its antlers. "Unless thou turnest to the Lord," a heavenly voice intones, to the bewilderment of the stag's pursuer, "and leadest an holy life, thou shalt quickly go down into Hell."

Hubert bows before the creature that a moment ago was his prey. "Lord," he asks, "what wouldst Thou have me do?"

It is a startling cross-cultural transformation: this is the myth of

Actaeon, overhauled to serve a distinctly different cosmology. Again we have a hunter, surprised to find himself in the company of a deity. Again a stag is imbued supernaturally with consciousness. But the resolutions of these two brushes with divinity, Actaeon's and Hubert's, could not be more divergent. Vengeful Diana chooses to make Actaeon the stag, thereby condemning him to a senseless, though symbolically appropriate, death in his own hounds' jaws. But the monotheistic deity, resonant with the divine sacrifice of the Christ story, makes Himself into the hunted form. To the early Christian mind, the Hubert tale also resounded because it played on both sides of a sometimes contradictory medieval fascination with the hunt. When hunting appears in medieval narratives, as the historian John Cummins notes, it generally "detaches a man . . . from his normal environment and, frequently, his companions, and takes him into unfamiliar territory"—territory that "is not merely topographical, but emotional and sometimes moral." In popular romance, it was in pursuit of the stag that heroes proved themselves above all.

Yet the finest stag was as revered as (or perhaps more revered than) the finest hunter was. The most exalted prey in the medieval hunt, the hart was believed to be uniquely holy. A stag pursued by hounds would sometimes figure as a marginal illustration in Bibles to symbolize good encroached by evil; one Christian allegorist likened the ten points of its antlers to the Ten Commandments. Bestiaries, in their treatment of the stag, would sometimes invoke Psalm 42: "As the hart panteth after the water brooks, so panteth my soul after you, O God." The devil, meanwhile, was portrayed as a hunter, setting traps for his human prey. One fourteenth-century German work goes so far as to say that Christ himself was hunted down and killed by "the hounds of hell and the infernal huntsman, the devil."

As that last example suggests, the dog was not seen in nearly so rosy a light. Though bestiaries often did remark on the helpful characteristics of dogs, they also lingered on not especially flattering examples: for instance, comparing the recalcitrant sinner to the dog that returns to its own vomit. In general, the spread of Christianity from the fourth to the eighth century had brought along with it a more uniformly dark vision of the dog, a view that is literally inscribed in scripture: of the forty-odd times that "dog" or "dogs" appears in the Bible, both Old and New Testaments, depictions of the creature range from revolting to merely distasteful. The best that the Bible can deign to say of a dog is this characteristically sardonic aphorism from the narrator of Ecclesiastes: "Anyone who is among the living has hope—even a live dog is better off than a dead lion!"

From there, though, it just gets nastier. As a holy people, the Israelites are ordered not to eat the flesh of wild beasts; instead, "throw it to the dogs." Dogs appear often as eaters of human flesh and drinkers of human blood. In 1 Kings alone, they are not just devouring corpses in the towns of Jeroboam and Baasha but also licking up the blood of Naboth; eating the flesh of his infamous murderess, Jezebel; and drinking the blood of her husband, Ahab, king of Israel. Five of the Psalms mention dogs, all painting the creatures as malign forces encroaching: for example, "Dogs have surrounded me; a band of evil men has encircled me, they have pierced my hands and my feet," or "They return at evening, snarling like dogs, and prowl about the city." The New Testament is hardly better. Our famous expression "pearls before swine" could just as easily have referred to man's best friend, given the original verse from Matthew: "Do not give dogs what is sacred; do not throw your pearls to pigs. If you do, they may trample them under their feet, and then turn and tear you to pieces." (It wasn't the pigs that would tear you to pieces.) In Luke, the ultimate insult to the beggar Lazarus is the dogs that come to lick his sores. In Philippians, Paul enjoins his audience to watch out for "those dogs, those men who do evil, those mutilators of the flesh." Even trippy Revelation gets in a parting shot, as the angel uses "the dogs" to lump together all those who will be left out come Judgment Day: "Outside are the dogs, those who practice magic arts, the sexually immoral, the murderers, the idolaters and everyone who loves and practices falsehood."

So in the medieval era, while popular storytelling lionized the hunter, the church's symbology denigrated both him and his devoted canine huntmates; indeed, it sanctified the very animal they most avidly sought to kill. The genius of the Hubert myth is in its clever fusion of these two somewhat opposed views. Hubert, the noble-born huntsman who finds glory and mystery in the chase, is the hero, and yet the particular form of his glory is in his submission to the stag, rather than the other way around. As the huntsman to whom God revealed Himself in the chase, he becomes, in the saint-crazed cults of medieval Europe, a supernatural master of the hunt and a guardian against the most savage spirit of the dog, as incarnated in the rabid bite. The myth concludes with Hubert leaving the hunting life and entering the priesthood, after the stag tells him to call upon the local bishop. "Go and seek Lambert," boomed the heavenly voice, "and he will instruct you."

Now, as for that elusive truth. We do know a few stray facts. There was a Hubert, and there was a Lambert. The latter, at the time of Hubert's supposed conversion, was the bishop of Maastricht, in what is now the Netherlands. We also know that in roughly the year 700, while on a trip to the nearby town of Liège, Lambert was murdered, though there are two differing and equally implausible accounts that describe the precise manner of his death. One places the doomed Lambert in a villa, lying on the floor during his murder, arms extended portentously in the position of the cross; another has him at the altar, with the assassin hurling a javelin from the congregation and piercing his heart. Here, of course, we are almost certainly back in the realm of myth. (As for the fate of the murderer, who both accounts agree was a miscreant called Dodo, we are assured that divine justice was soon done, as "his hidden parts were made rotten and stinking" and then "cast forth through his mouth.")

Consequently elevated to the position of bishop, the historical Hubert looked into Lambert's life and uncovered many miraculous doings. The pope soon agreed to a beatification. To house Lambert's body and effects, Hubert built a new church in Liège, the site of the martyrdom, and made it the seat of his bishopric. And after Hubert's own, significantly more peaceful demise in the year 727, his underlings performed a similar trick on his behalf: another inquiry, written up as an official hagiography, discovered Hubert's own set of otherworldly interventions. To a region stricken by drought, he had brought rain. In one town, he had doused a fire. He had healed the afflicted—though none from rabies as of yet. Most dramatically, as monks had discovered some time after his death, Hubert's very body had failed to decay: "Briefly approaching the tomb with great fear, beholding a light from within, they discovered his glorious body in the tomb solid and incorrupt." The corpse even emitted a "miraculously sweet smell."

As Hubert himself had proved with the transfer of Lambert's relics to Liège, the remains of a new saint represented a sort of spiritual and economic windfall. Some decades after Hubert's death, a struggling abbey in the village of Andage petitioned the current bishop, Waltcaud, to allow it to obtain Hubert's body and holy effects. Some three years elapsed before the bishop was able to bring the question to the Frankish emperor, Louis the Pious; he, in turn, left the question to the local synod, which eventually gave its assent. In the year 825, Hubert's remains were brought to Andage, where the abbey, and soon the town, were renamed Saint-Hubert in his honor. The saint's cult quickly grew, and with steady donations from seven centuries of sufferers and supplicants the abbey grew, too; after it burned in 1525, it was rebuilt on an even grander scale throughout the remainder of the sixteenth century. In 1607, the abbey's hospice, which succored those patients for whom prayer did not suffice, was relocated below the abbey to a new building.

Today, more than five hundred years later, that former hospice houses an upscale hotel and wine bar, called L'Ancien Hôpital, whose friendly young hoteliers, Hans and Ann Swaan–Van Tilborg, live there with their young son, Andreas. The hospital's original seventeenth-century chapel remains intact, now a somewhat gloomy but (we can attest) rather comfortable guest room where guests can commune with the spirits of agonized hydrophobes while soaking in a Jacuzzi tub. One can re-create the journey to Saint-Hubert in considerably more comfort, too, with high-speed rail available for most of the journey. The trip does require a switch to a local in Liège, a drab burg whose ecclesiastical roots gave way to heavy industry in the nineteenth and twentieth centuries—when it became known for coal and steel—and whose fortunes in the twenty-first have been diminished by competition from China. Then, from the nearby town of Libramont, the way is made by a punctual local bus with a laconic driver. Upon arrival in Saint-Hubert, if the forty-five-euro prix fixe at the former hospital strikes one as too steep, one can fully gorge for just a few euros on *pommes frites* and curry ketchup from the snack shop down the street. Catty-corner is the soaring facade of the mammoth church that grew from Saint-Hubert's tiny abbey over the centuries, a behemoth that during the 1920s was elevated by Pope Pius XI to the rank of *Basilica minor.*

Inside the church, visitors are greeted by a compact and serene docent of indeterminable age, Marie-Françoise Rakotovao. A native of Madagascar, Rakotovao speaks French, is learning Dutch, and is kind enough to give an English-language tour to two quizzical Americans. Though pilgrims no longer come for hydrophobia cures, the docent waxes passionate about how, at least in a metaphorical way, the healing power of Saint Hubert remains relevant to the contemporary seeker. "We have our own rabies *here,*" she proclaims, clutching a hand to her chest. "We are depressed. Something is wrong in our hearts. Something is wrong in our *minds.*"

Throughout the basilica, in stone and enamel and wood, countless elaborate depictions attest to Hubert's conversion legend. At the Basilica of Saint Hubert, the stag rears his holy head everywhere: carved in stone on high pedestals, next to the praying angels; in a series of meticulous woodcuts in the choir, laying out the full legend; on a statue of Hubert on the church's very facade, peering out from behind his robes, small and docile and doglike. A full hunting party appears in a ravishing oil painting that dominates the nave, the saint's dogs cowering in confusion as the saint prostrates himself before the noble buck. The entire Hubert myth, it should be mentioned, was not original to him: it was pilfered wholesale from the almost identical legend of Saint Eustace, a former Roman general who lived in the first century A.D. (Eustace suffered enough in life, if said legend holds: he and his entire family were roasted to death inside a bronze statue of a bull.)

Perhaps it is fitting, then—karmic, even, if we may borrow from a different creed—that so much would be quite literally stolen from Saint Hubert in the centuries after his death. Visitors can tour the crypt where his relics were once kept, but these disappeared long ago, in 1568, when the abbey was raided by the Huguenots. In the worst blow of all, the marauders also made off with Saint Hubert himself—a barbaric theft of not just the abbey's soul but its livelihood.* In the absence of the body, attention shifted to his sacred vestment, which dates from the twelfth century (that is, four hundred years after the actual saint). Overlooked somehow by the rampaging Huguenots, the moldering stole now sits alone in a gilded display case, the centerpiece of a modest reliquary in the church's southern transept.

Most crucial to our own quest is an unassuming metal ring, affixed to the wall across from the vestment case, between two red padded benches. This was the site of *la taille,* the holy rabies treatment. From a white paper bag, the type that might normally cradle a croissant, our guide shakes out some original instruments of this treatment—a broad, dull scalpel, a metal nail the size of a golf tee—the sight of which spurs, in a visitor, the sort of sinking terror one imagines a patient felt at this point in his pilgrimage. Bound to the sturdy ring in the wall,

---

* This travesty did not go unnoticed in the church's reconstruction. The gilt frame of one enameled display bears a chronogram, or a Latin message carrying a date inside it, reading: "*ConCVLCaVerVntsanCtIfICatIoneM*"—or, "They have spurned that which is holy," with the numerals spelling out 1568.

presumably to prevent a last-minute change of plans, the patient was slashed across the forehead, and in this wound was placed a thread from the venerated vestment. The wound was then bound for nine days, during which time the patient remained in the abbey, praying and fasting, dressing all in white. Combing of the hair was strictly forbidden. On the tenth day, a priest removed the bandage and burned it.

During our visit, we asked Rakotovao when the last administration of *la taille* occurred at the basilica. The answer: 1919, or some four decades after the invention of the rabies vaccine. Contemplating the majesty all around, one can perhaps understand why. No set of shots in a drab doctor's office could infuse the prevention of rabies with such a connection to history, to the natural world, to the symbology of faith. *La taille* was literally a form of communion, except that the blood poured was one's own—shed so that the sins of the dog might be forgiven.

If there was a dualism of the dog in medieval Europe, it broke down not within each dog but between rich owners and poor. The nobility carried on the Greek veneration of the well-bred dog, and in particular the hunting hound. The French aristocrat Gaston III, Count of Foix, writes in his widely read (and imitated) hunting book *Livre de chasse*—written circa 1388—about the ideal running hound, the *chien baut,* in which commingle all the finest canine attributes: not just beauty and obedience, but a nearly supernatural ability to track prey and to communicate with human masters. "The *chien baut* must not give up on its beast, not for rain nor wind nor heat nor any other weather," writes Gaston, "and it must hunt its beast all day without the aid of man, just as if man were with it always." (Gaston said he had encountered only three *chiens bauts* during his long life of hunting.) The Castilian king Alfonso XI recommended that the finest alaunts, a medieval hunting breed closely related to the mastiff, be allowed to live in the palace; one Portuguese prince of the fourteenth century so loved his two alaunts, Bravor and Rebez, that they slept on either side of him in his bed. Devotion to the hunt extended to the clergy as well. One medieval archbishop of Canterbury kept twenty hunting grounds of his own, while even Thomas à Becket, when serving as Henry II's ambassador to France, insisted that hunting dogs accompany him in his retinue.

Medieval love of the hound was not entirely a masculine province. Among the best-loved English treatises on hunting is the fifteenth-century *Boke of Saint Albans,* written by one Juliana Berners, prioress of the Sopwell nunnery. Penned entirely in verse, Berners's book limns the look of the ideal greyhound: "A grehounde sholde be heeded lyke a snake: and neckyd lyke a drake: fotyd lyke a catte: tayllyd lyke a ratte,"* and so on. Indeed, the highborn woman's devotion to hounds—and vice versa—is a common trope in accounts of medieval life, particularly regarding those women who, like Berners, resided in convents. Chaucer's own Prioress, in *The Canterbury Tales,* travels with "small houndes" that she "fedde with rosted flesh, or milk and wastel-breed," that is, white bread. (Such a fondness for dogs among nuns is amply documented in real life, too. In 1387, at roughly the same time that Chaucer was writing his "Prioress's Tale," the bishop William of Wykeham upbraided one particular abbey in stark terms, noting that "the alms that should be given to the poor are devoured," and the church itself "foully defiled," by the "hunting dogs and other hounds" in residence at the abbey. Therefore, he went on, "we strictly command and enjoin you, Lady Abbess, that you remove the dogs altogether and that you suffer them never henceforth, not any such hounds, to abide within the precincts of your nunnery.")†

For the poor, though, the dog took on a decidedly different cast of meaning. Medieval towns and cities were no less amenable to the

* That is, the hound should possess a head like that of a snake, a neck like that of a duck, feet like those of a cat, a tail like that of a rat, and so on. This, it should humbly be noted, perfectly describes our own dog Mia, a whippet.

† More morbidly, we have the story, handed down regarding the death of Mary, Queen of Scots (1587), that one of her executioners, while removing the garters from her corpse, "espied her little dogg which was crept under her clothes," a poor creature that eventually "came and lay betweene her head and her shoulders."

scourge of semi-wild dogs than were their Roman, Greek, Egyptian, or even Mesopotamian precursors. Some of the poor did own dogs, and occasionally kept them in houses, but these likely were seen as working animals, the cost of whose nourishment (which mattered dearly in lives lived so close to bone) had to be offset through useful farm labor on their parts. A dog could be a terrible liability in a feudal society: peasants whose dogs romped in the wrong forest could incur fines, as seen in this list of English public records from the reign of Edward II: "From John de Maunchestre for one dog, 3s. From Wilto le Seriaunte for one dog, 3s. . . . From Wilto de Huntyngtone for one dog, because he was poor, 12d." In other accounts, peasants who took game from the preserves of their betters found themselves blinded, castrated, or even killed. Such draconian enforcement of the hunt as a noble privilege extended so far as to include preemptive rules about dogs. It was standard practice for all commoners' dogs living near the royal hunting forest to be "expeditated," that is, rendered unable to run by having one or more claws hacked from a foot. Peasants could not even legally own greyhounds in England, a prohibition that dated to the eleventh century.

Should anyone doubt the dark heart of the dog, as revealed to medieval minds, he can find particularly vivid testimony in accounts of the Black Death, which ravaged three continents between 1347 and 1350, culling (by some estimates) more than half the population of Europe and then returning intermittently for centuries. These devastating epidemics transformed half-tamed neighborhood dogs into demonic corpse eaters. Agnolo di Tura, a shoemaker in Siena, Italy, recounted seeing, during the 1347 outbreak there, "many dead throughout the city who were so sparsely covered with earth that the dogs dragged them forth and devoured their bodies." About a 1429 resurgence in Cairo, one chronicle reports grave diggers carving out giant trenches into which bodies were slowly heaped, while dogs fed on their outer extremities. A Florentine priest described, during a 1630 plague, bodies thrown in piles

> as if they were mounds of hay or piles of wood; only, I say, if they had been hay or wood they would have been stacked more neatly; but they will be stacked there haphazardly, some half covered, some with an arm exposed, some with their head and some with their feet left as prey for the meals of dogs and other beasts.

In painted plague scenes, dogs nosing at the corpses of the dead became a stock trope; the expression "six feet under" originated from a London health ordinance during the plague of 1665 there, with the famous prescription intended to keep men from being unearthed by man's best friend.

As we know today, the pathogen that causes bubonic plague, *Yersinia pestis,* is a particularly deadly zoonosis, which persisted in its decimation of human populations only through its reservoirs in the rat. (Today *Y. pestis* is also known to crop up sporadically from marmots and prairie dogs.) Indeed, the animal nature of the infection was crucial to its unprecedented rates of fatality. In humans alone, an epidemic that wipes out more than a third of its victims will tend to snuff itself out for lack of new hosts, or at least evolve into a less virulent strain. But the rats—which do die from the disease, but at a far lower rate than humans—could scurry from town to town with impunity.

From the pathogen's perspective, humans are what epidemiologists call an "accidental host" of many zoonoses, meaning that the pathogen usually fails to complete its life cycle in man alone. In other words, it can "afford" (evolutionarily speaking) to kill humans at staggering rates, because its natural reservoir is elsewhere. Rabies' residence in people is also, by these standards, accidental, though its inability to spread through humans largely boils down to issues of anatomy and behavior: although the virus does express itself in human saliva, humans lack a propensity to bite and the sharpened teeth with which to do it effectively. In the case of plague, the bacterium has evolved to pass most efficiently from rat to rat through a flea's digestive tract. The millions of vulnerable humans whose skins those fleas afflicted were history's largest class of collateral victims.

Medieval scholars, medical and otherwise, had no inkling that the plague was originating in their rodent neighbors. The preeminent Arab physician Ibn Sīna does make an intriguing offhand remark, in his *Canon of Medicine,* about how a sign of pestilence is the emergence of mice and other animals, who run about as if inebriated. But in 1349, one of his intellectual heirs, Ibn Khātimah, wrote a tract theorizing that the new and terrible worldwide plague was essentially caused by bad air. He believed that in those instances where thousands died in a single day, the air had become entirely corrupted such that it was almost a different substance, like in wells where dead animals have been cooped up to decay.

The explanations of Christian scholars and physicians were roughly comparable. Gentile of Foligno thought the bad air entered victims through their wide pores—the wider the pores, the more susceptible the person—and was then drawn into the heart. A tractate from the medical faculty of the University of Paris held that the air had been corrupted by noxious vapors, brought on by the movement of the planets but exacerbated by the southerly winds of late. Alfonso of Córdoba likewise blamed astronomic happenings for the plague's onset but felt that its continuing spread was due to a few very crafty individuals, taking out their enemies:

> The person who wishes to do that evil waits till there is a strong slow wind from some region of the world, then goes against the wind, and puts his flask [full of infected air] against the rocks opposite the city or town which he wishes to infect, and making a wide detour by going back against the wind lest the vapor infect him, pulls his flask violently over the rocks.

A number of the fourteenth-century authors did comprehend the idea of person-to-person infection, though its mechanism was obscure to them: one, a physician from the French town of Montpellier, held that the disease was passed through the optic nerves, such that the well man might look at the sick and immediately be attacked by the pestilence.

No link between the plague and the rats, let alone their fleas, could be apprehended by anyone at the time. Indeed, with the exception of rabies, the whole notion of zoonosis, that humans were being infected by animals, was almost entirely foreign to the medical mind of the Middle Ages. The only other zoonotic disease that seems to have been understood at the time was anthrax—the cattle disease that causes both skin lesions and (if inhaled) death in humans. But awareness of anthrax over the ages was dim and intermittent. While the Bible describes an anthrax-like plague in Exodus 9 ("festering boils . . . break out on people and animals throughout the land") and a sixteenth-century "boke of husbandry" describes a similar malady, no descriptions of the disease are to be found in the intervening medical scholarship. (The term *anthrakes* dates back to Hippocrates, but it is used to describe not the zoonotic disease but the black carbuncles on human skin that typify a number of diseases, most probably smallpox.)

Yet increasing urbanization, and more intensive use of agriculture, were accelerating the rate at which infections passed from animals to humans. And by the fifteenth century a third factor—man's new mobility, carried out in ocean voyages of staggering length—brought much of the world's people into contact with devastating germs they had never before encountered. This so-called age of discovery saw two European neighbors, Spain and Portugal, amass enormous empires as they pursued wealth from expanding trade in spices and precious metals; Spain would seize far more territory, colonizing most of the Americas, while Portugal would cover the globe more finely, setting up outposts from Brazil to Angola to Goa to Macau. In ferrying goods, these colonists also ferried germs in all directions.

The bulk of these were human-to-human infections. Most

notably, they brought smallpox to the Americas but took away syphilis, which tore through Europe and then eastward into Asia. But meeting new animals meant meeting their diseases, too.

It's arguable that the greatest devastation Christopher Columbus wrought upon the New World took the form of eight sows, which made landfall in Hispaniola on December 8, 1493. Those pigs are believed to have instigated a massive swine flu epidemic that began the following day, killing Indians in staggering number—"the stench was very great and pestiferous," observed one Spanish civil servant soberly—and initiating a long run of pestilence that offed some two-thirds of Santo Domingo's natives in little more than a decade. One Dominican friar later put the death toll from disease in Hispaniola at nearly a million, in less than thirty years, and described similar die-offs in Cuba, Jamaica, and Puerto Rico.

But at the time, no one seemed to have looked askance at those eight fateful swine.* Columbus himself, who fell ill from the flu early in its run and took some three months to recover, was an especially poor epidemiologist: he wrote home that "the cause of the ailments so common among us, is the sustenance, and the waters, and airs."

Rabies, in contrast to other zoonoses, was universally known and widely feared. Keeping hounds free from its ravages was a key preoccupation of the medieval huntsman. Edward of Norwich, the second Duke of York, who was born in 1373 and died in 1415 at the Battle of Agincourt, published an English-language translation and expansion—entitled *The Master of Game*—of Gaston's famous hunting book. It includes a chapter called "Of Sicknesses of Hounds and of Their Corruptions," tucked in just before the languorous celebrations of the various hound breeds but after the elucidations of their

* In our most recent outbreak of swine flu, the pigs were not so lucky, particularly in the Muslim world: see Chapter 6.

various prey (hare, hart, buck, roe, boar, wolf, fox, badger, wildcat, and, last but certainly not least, otter). The first dog malady, and the one considered at greatest length, is the "furious madness":

> The hounds that be mad of that madness cry and howl with a loud voice, and not in the way that they were wont to when they were in health. When they escape they go everywhere biting both men and women and all that they find before them. And they have a wonderful perilous biting, for if they bite anything, with great pain it shall escape thereof if they draw blood, that it shall go mad whatever thing it be.

It is recognized that such a condition, if not temporary, is invariably fatal: "Their madness cannot last but nine days but they shall never be whole but dead." As a protective measure against rabies, the book follows Pliny in recommending that the "worm" beneath a hound's tongue—in truth, a ligament—be cut out, by disabling the jaws with a staff and cutting out the offending worm with needle and thread or (in Edward's addition) "a small pin of wood."

For men bitten by mad dogs, *The Master of Game* suggests a number of remedies, though a few of these are dismissed in the same breath as they are proposed. For example, some bitten men go to the sea and allow nine waves to pass over them, but "that is but of little help." Other men pull all the feathers from around a live rooster's anus and, hanging the poor bird by the neck and wings, set the anus on the bite wound, on the theory that said anus would suck forth the poison. If the rooster swells up and dies, then the hound is mad, but the man will be healed; that is, the book avers, "many men say" this is the case, but "thereof I make no affirmation." The authors seem to feel on more solid footing with the suggestions of cauterization ("it is a good thing for to hollow it all about the biting with a hot iron") and bloodletting, both of which might actually have done the victim some good. They also spell out some recipes for medicinal salves. First is prescribed a sauce of salt, vinegar, garlic, and nettles; second, and more appetizing still, is a paste in which the garlic and nettles have been combined with leeks, chives, olive oil, and vinegar.*

The most lavish prophylaxis against hydrophobia in the hunting hound was carried out, fittingly, by the kings of France. In the hunting accounts of the French palace, historians have found annual outlays for all the king's hounds to undergo a special ceremony. They were transported to the Church of St. Menier les Moret, in order "to have a mass sung in the presence of the said hounds, and to offer candles in their sight, for fear of the *mal de rage*"—that is, the disease of rabies. One wonders whether the hounds howled along.

The medieval period is also when our contemporary terms for rabies and hydrophobia began to enter the vernacular—in both their literal and metaphorical meanings. The French word *rage* is a derivative of *rabies,* which in Latin served as a rough equivalent of *lyssa.* As with that Greek term, *rage* in French begins its life both as a horrible disease and as something more profound, a sort of animalistic fury tinged with madness. The earliest documented uses are in *The Song of Roland,* circa 1100; the first of these appears in the reaction by the king, upon hearing that the treacherous count Ganelon has tapped Roland, his own stepson, to lead the rear guard (Ganelon has cut a deal with the enemy, the Muslim king Marsil in Spain, that his Saracens will attack from the rear). Says the poet:

> *When the King hears, he looks upon him straight,*
> *And says to him: "You devil incarnate;*
> *Into your heart is come a mortal hate [rage]."*

---

* Another gastronomic treatment is supplied by *Le ménagier de Paris,* a fourteenth-century guide to domestic life that prescribes, at the end of a long list of recipes, a novel treatment for rabid dog bite. "Take a crust of bread," it advises, "and write what follows: *Bestera bestie nay brigonay dictera sagragan es domina siat siat siat.*"

The second is used to describe the folly of a Saracen who, in the heat of battle, launches an ill-advised attack on Roland: "*Par sun orgoill cumencet mortel rage,*" or "A mortal hate he's kindled in his pride."

It was not until the seventeenth century that "rabies" and "rabid" seem to have found purchase in English. Interestingly, the *OED*'s earliest use of the latter precedes the former—a reference to "rabid mastifs" in a 1596 translation from the Greek. The seventeenth-century uses of both terms seem to have been restricted, if not to the literal disease (or the semblance of it) in animals, then to a particularly vociferous rage, verging on madness: for example, "Rabid with anguish, he retorts his looke Vpon the wound" (1621); "Hee . . . strokes and tames my rabid Griefe" (1646). But rabies is irresistible as a metaphor; in both English and French, the word began to fan out into far more playful contexts within a century or two of adoption in both cases. By 1288, a French wag is wondering over "*tel conseil et tel rage,*" or "such counsel and such rage," that is given to the king; by 1678, we have the very contemporary sense, true in English as well, of rage as a fashion, for example, *la rage de la bassette,* the faddishness of a then-popular card game. "Rabid" in English being significantly newer, the drift occurred somewhat later, and so we must wait until the nineteenth century to get, for example, the "rabid desire for the good opinion of every thing human" (1838).

How should we feel about these uses of rabies as simile, as trope, as joke? In *Illness as Metaphor,* Susan Sontag chronicled the myriad ways that our popular understandings of disease—tuberculosis and cancer, in particular—have been polluted by literary associations throughout history. As she demonstrates, disease metaphors gave rise to a particularly pernicious form of pseudoscience, fostering the myth that certain personality types (the romantic, decadent figure in the case of TB, the repressed and dissatisfied in the case of cancer) had a greater propensity to contract each disease. Metaphors also helped to stigmatize the ill, as moral judgment slipped backward from literary invocations of the disease ("cancer in the body politic," or Victor Hugo's remark in *Les misérables* that monasticism is "for civilization a sort of tuberculosis") to taint the sufferer. Sontag's point, which she declares with thunder on her very first page, "is that illness is *not* a metaphor, and that the most truthful way of regarding illness—and the healthiest way of being ill—is one most purified of, most resistant to, metaphoric thinking."

The injunction is an admirable one, but it's hard to look at the function of disease in history (especially in history before the Pasteurian era of the late nineteenth century, when people first learned about the microscopic nature of disease) without concluding the injunction to be an impracticable one as well. For millennia, disease *was* metaphor—the carrying of a mysterious foreign meaning within the familiar vessel of a human being; the very etymology of the term (*pherō,* or "carry," plus *meta,* or "across") invoked just such an act of freightage. And this inevitability of metaphor in disease is nowhere more present than with rabies, where the name itself in multiple languages—*lyssa, rabies, rage, rabia*—also describes a human emotion of fury, with the twinned meaning extending back indefinitely, neither the medical nor the figurative sense taking clear precedence. Rabies was *identical* with a visitation of animal rage; or, if it was not quite a true identity between the two, the link transcended mere metaphor to become intrinsic to both poles of the comparison. Insofar as rabies (from the Babylonians to *The Office*) has served as grist for comedy, it's the extremity of the fury that provides the punch line—the rage for the card game, or the "rabid desire" for others' acclaim, or being a "rabid Justin Bieber fan," all derive their implicit wit from the notion of aficionados so passionate as to be foaming at the mouth.

During the medieval period, the real advances in the medical understanding of rabies—as with advances in medicine more generally, for that matter—occurred almost entirely in the Islamic world. Through assiduous translation to Arabic, the great Muslim physicians perpetuated and extended the work of the Greeks and Romans. Arguably this process began in fifth-century Syria, where fleeing Christian heretics, expelled from the Byzantine church after denying the Nicene Creed, commingled with Arab-speaking Muslims and brought Greek traditions and texts to their attention. By the tenth century, Baghdad boasted an impressive system of hospitals; the thirteenth century saw the establishment of the Arab world's first medical madrassa, in the old goldsmiths' quarter of Damascus. Medieval Islamic physicians are even believed to have been the first to lay out a process akin to scholarly peer review: as prescribed by the Syrian doctor al-Ruhāwī, cases were double-checked afterward by a local council of fellow physicians, on the basis of whose judgment a doctor could be sued for malpractice.

The three titans of medieval Islamic medicine—al-Rāzī (known to European history as Rhazes), Ibn Sīna (Avicenna), and Ibn Zuhr (Avenzoar)—all addressed rabies in some detail in their central works. The first of these scholars, al-Rāzī, who wrote and practiced in Baghdad at the turn of the tenth century, recounted the cases of rabies he had personally witnessed:

> There was with us in hospital one such man who barked during the night and then died. Another did not drink water, but when some water was brought to him, he was not afraid of it, but said: "It stinks, and the stomachs of dogs and cats are in it." Yet another patient, when he saw water, shuddered, shivered, and trembled until it was taken away from him.

His preferred treatment for bites—cauterizing and scarifying the wound, followed by the application of suction cups to it—is as sensible as any described to that time.

Ibn Zuhr, who wrote in Spain between 1121 and 1162, included an essay called "On Furious Madness" in the *Kitab al-Taysîr,* his magnum opus. As al-Rāzī did before him, he anchors his observation with a personal narrative. "My father," he wrote, "to whom God grants mercy, taught me that a mule, having been affected by this sickness, will aim to bite a man. The latter, fleeing before the animal, will enter into an alleyway whose entrance is quite narrow. The mule will rush towards the entrance, head down, and will become so tightly compressed in it that it will only be able to extricate itself by its own death."

The sentiments on rabies expounded by Ibn Sīna, the most heralded of all medieval Islamic authors, seem lamentably off-key by today's standards, even when compared with the views of his contemporaries. In the fourth book of his mammoth *Al-qanun fi al-tibb* (*The Canon of Medicine*)—composed during the early eleventh century, in Persia—the great doctor expressed the belief that heat and cold helped foment the disease, causing this "serious and venomous melancholy" in dogs. Moreover, he attributed it to the consumption of bad water and bad meat. Among the symptoms Ibn Sīna describes in human patients is a hallucination of little dogs, which one supposes to be possible enough. But among the treatments he recommends for human sufferers is cantharis, the legendary aphrodisiac that today we call Spanish fly; this seems a particularly odd choice for a malady whose symptoms, even on Ibn Sīna's list, include priapism.

A fairly lengthy treatment of rabies is found in the writings of Moses Maimonides, a twelfth-century Jewish philosopher and physician who practiced in Morocco and then, later, in Egypt. He recognized, contrary to widely held belief, that the bite of a mad dog did not cause greater pain than that of a healthy one. More important, he states outright that any treatments for the rabid bite are useless after the onset of hydrophobia. He also recognized that symptoms of madness in humans can often be delayed by a month or more. Besides the time-honored treatments, which he endorses in passing—widening the bite through incision, bloodletting through cupping, and so on—he prescribes a dizzying array of potions and poultices: the pulverized ashes of river crabs, drunk in water daily; bitter almonds crushed in honey and applied to the affected site; a raw bean, chewed into a paste and then rubbed on the wounds; or, by the same method, crushed wheat, or onions, or unleavened bread. Maimonides wisely urges caution on bites, instructing readers to fear the worst. "If the condition of the dog is in doubt," he writes, "conduct yourself as if the dog is mad."

Artifact collectors have preserved one last rabies treatment from the era, one that seems to have begun in the twelfth century: the magic-medicinal bowl, a metal vessel elaborately inscribed with various therapeutic instructions, such as hot water to aid colic or saffron water to ease a difficult childbirth. A cure for rabies, and for animal poisons more generally, was often promised by these bowls. The oldest surviving specimen, made in 1167 for the Syrian ruler Nūr al-Dīn Mahmūd ibn Zangī, claims to assuage not just the rabid bite (through the drinking of milk, water, or oil, "by the help of God Almighty") but also chest pain, migraine, and even demonic possession.

During the early days of the Inquisition, in the latter decades of the fifteenth century, a mysterious brotherhood of healers roamed from town to town, offering protection against rabies. These were the so-called *saludadores,* and they crackled (so they claimed) with powers bestowed by hallowed saints. They were blood relatives of Saint Catherine of Alexandria and carried her mark: a toothed wheel, representing the apparatus on which the third-century saint was tortured to death. Or they were in league with Saint Quiteria, another early Christian martyr whose intercession was often called upon for protection from rabies. (Quiteria also had the distinction, in Portuguese Christian lore, of being the leader of a savage all-girl team of nonuplet infidel slayers.) Through the grace of these sanctified women, *saludadores* could nullify the savage bite, often with their saliva or their breath. They could touch red-hot iron, wash their hands with boiling oil, or even clamber inside a fiery oven without injury.

The Inquisition saw such claims as heretical, and its official position was to squelch the *saludadores.* The few firsthand accounts from these healers that survive tend to be from those who confessed, under questioning, to having been frauds. In 1619, a shoemaker named Gabriel Monteche confessed that he had

> held the office of *saludador* for many years, pretending he had the virtue to cure the bites of rabid dogs, and to cure other sicknesses and to deliver villages from hailstorms, saying that he bore on one arm the wheel of Saint Catherine and on the other a cross, which signs he had made himself with a needle to deceive people and let them think he had been born with them.

He went on to describe how he snookered the marks:

> He would put a worm in his mouth, and let it be thought by those who had been touched by rabid dogs that he was a *saludador* and that he would heal them. And he would have a surgeon pierce the patient's skin, allowing a little blood to spill, and then he would come, suck that blood, and afterwards add it to a bowl of water and, having stirred the two, would add the worm from his mouth, and as it was mixed with the blood that he had sucked, they thought and believed that he had taken it from the man's body.

Chicanery like this, if indeed this confession was honest and not forced, was probably anomalous: folk healing never sustains itself entirely on the basis of bad faith, and no doubt the majority believed in their powers every bit as fervently as did their patients. By that time, medicine in Spain and Portugal had become licensed, normalized—from the physicians in official hospitals, which sprouted up over the sixteenth century, all the way down to the lowly "barber-surgeons," who both shaved and operated on customers. But then as now, for reasons of cost and of idiosyncratic superstitious belief, many patients preferred unofficial practitioners such as *saludadores*. It did not hurt, of course, that in many cases the official medicine was no more effective. Certainly this was the case with rabies, which was no more curable (or even preventable) than it had been in the second century A.D.

Many Spaniards preferred their religion unofficial, too. We tend to think of the Inquisition as a totalitarian regime, at least on spiritual matters, but in fact it did little to temper the riot of local eccentricities across the lands it ruled. On two occasions during the 1570s, a Spanish royal office made a survey throughout the kingdom, asking two or more representatives from each town to answer a series of questions about its population and practices. Some of the questions involved religious belief: respondents were asked to detail the chapels in the town, the miracles that had taken place there, the holy and fast days observed there. The office wound up surveying 513 towns, representing a little more than 127,000 households; and the replies enumerated a remarkable diversity of religious practice. On the question of holy days, locally observed as part of a vow to some saint, there were more than fourteen hundred different vows spelled out, made to dozens of sactified figures. Residents of Cabezarados, which sits roughly halfway between Madrid and Córdoba in the outskirts of Ciudad Real, reported that they had recently lapsed in their vow to Saint Quiteria, causing a rabid wolf to kill a young man and bite a number of cows. "Since these events," the royal chroniclers later wrote, "the townspeople have observed and do observe the old vow with much devotion and hold a solemn procession and feed all the poor in the town; and everyone from the town eats in the house of the mayordomo that day, each paying his share."

In practice, the Inquisition in Spain took a stance toward the *saludadores* that one might call benign neglect. One intriguing reason for this, as the Spanish historian María Tausiet has documented, is that *saludadores* also had a reputation as crackerjack witch-hunters. Documents from the era show that many of them, despite their so-called marks of the devil, worked closely with both the inquisitorial and the secular justice systems in identifying witches. It was not uncommon for investigators to bring a *saludador* along with them as they swept into town. Inquisition records note that a healer named Andrés Mascarón condemned thirteen women in the village of Bielsa as witches, saying that "on seeing a witch he felt his flesh burn, and the older the witch the more it burned." Four of these women were summarily hanged, and the rest sent into exile; the town paid him generously for his efforts.

One can infer, through this strange dual role of the *saludador,* the demonic nature of rabies as it was perceived at the time. The healer was the exorcist of hydrophobia, the diviner of witchcraft; he was, that is, the enemy of all the malign animal spirits that seized unawares the innocent human soul. So many of the superstitions that ran roughshod through the fevered medieval imagination had what was, at base, an animal element. In the next chapter, we chart the zoonotic idea as it manifested itself in two enduring terrors of bestial infection: the werewolf and the vampire.

The Werewolf by Lucas Cranach, c. 1510–15.

3

A VIRUS WITH TEETH?

In September 1998, the journal *Neurology*—which usually consumes its column inches with such thrilling topics as "detection of elevated levels of α-synuclein oligomers in CSF from patients with Parkinson disease"—gave voice instead to an eccentric theory on a historical conundrum. Over four densely cited pages, a Spanish physician named Juan Gómez-Alonso put forward the argument that rabies, a subject dear to neurologists for its uniquely devastating effects upon the brain, might also serve as an explanation for one of our oldest horrors: the vampire, whose roots stretch back to ancient Greece but whose alleged romps through eastern Europe during the eighteenth century launched a mass fascination that continues to this day.

Gómez-Alonso's hypothesis made headlines around the world, from Los Angeles to London to Sydney. Even *Playboy* weighed in, noting the doctor's linking of both rabies and vampirism to hypersexuality. "Bite me!" the writer enthused. It's easy to understand why the public's interest was so piqued. Our myth of bloodsucking ghouls has proven remarkably durable throughout the last two hundred years of churning popular culture, sinking its teeth into everything from Victorian novels and Hollywood confections to Anne Rice's wildly popular novels and, of course, the multiplatform tween juggernaut that is Stephenie Meyer's *Twilight series.* We feel we owe a hearing to any theory that might explain the origins of these surprisingly unkillable undead.

Gómez-Alonso's paper does raise many intriguing parallels between the vampire and the sufferer of hydrophobia. First, and most obvious, both rabies and vampirism spread from organism to organism through bites: not a small coincidence in man, an animal that does not instinctively use its teeth in aggression. Also, the throes of a rabies infection usually involve facial spasms, which can render an appearance—as a 1950 French medical text described it—of "the teeth clenched and the lips retracted as those of an animal." Vampires were believed to possess the ability to become dogs at will, and in this form they would often kill the other dogs nearby. Male rabies patients, as *Playboy* was so excited to learn, are sometimes given to undue sexual abandon; vampires, meanwhile, rose from their graves to engage in sexual conquest. And finally, the life span of a vampire was said to be forty days, similar to the average duration of human rabies infections from the time of bite until death.

The doctor points out in passing that rabies might also account for the werewolf, or lycanthrope, that mythical human who changes wholly or partially into a wolf and preys upon his neighbors. Gómez-Alonso does not provide specifics, but the broad strokes of the comparison are obvious: the biting, the clenching teeth, the animal transformation, are all even more pronounced in the myth of the wolf-man. The parallel to rabies is, if anything, even more direct with lycanthropy, which is nothing more nor less than a man seized with an animal nature.

How much credence should we give to the link between rabies and the undead? In his paper, Dr. Gómez-Alonso goes so far as to assert that the vampires and werewolves in historical accounts were *literally* rabid humans, their symptoms misunderstood as supernatural by an unscientific populace. In propounding this theory—in attempting to explain away folkloric evil through science—the doctor joined a noble tradition that extends back at least to Europe's great vampire boom, in the early part of the eighteenth century, when supposedly true-life tales of vampires from the East chilled the drawing rooms of England, Germany, and particularly France, where, as Voltaire famously wrote, "nothing was spoken of but vampires, from 1730 to 1735." This was the self-described age of reason, after all, and its eminent minds, such as Voltaire and Rousseau, brooded over how seemingly respectable people could display such credulity toward popular hysterias. Thus even during the vampire's heyday, men of reason gamely tried to offer scientific explanations; some saw vampires as victims of food poisoning or as opium fiends.

When one digs into historical accounts, however, such literalistic explanations seem far-fetched, to say the least. Werewolves, during the sixteenth century, were apprehended and would seem entirely lucid (and fully human) during what by accounts were lengthy interrogations and trials—not something that a rabies sufferer could have accomplished. In the accounts of vampires from the eighteenth century, observers would disinter real corpses that appeared, in the light of day, to be entirely dead, not writhing in any sort of rabic agony. And there is the unavoidable fact that rabies, for all the violence of its manifestation in humans, rarely prompts them to bite and also does not shed abundantly in their saliva as it does with dogs. Simply put, humans do not spread rabies.

Yet Dr. Gómez-Alonso's theory, if questionable in its literal meaning, taps into a deeper metaphorical truth. So many of our most enduring horrors, the vampire and the werewolf included, have common narrative elements that derive naturally (in both senses of that word) from rabies. Just browse the horror-movie section of your local video store and see what's on offer. It's villains pouncing from the darkness, biting, lunging, tearing with claws. It's contagion: a malevolence that creeps from victim to victim, spreading through bites, kisses, licks. It's a familiar creature—a trusted soul, often residing within one's own inner circle or even within one's home—that becomes surprisingly and unaccountably infected by a savage animal evil. Going as far back as the days of *lyssa,* and even before, these fiendish tropes have been forever intertwined with rabies, a constant presence across continents and across eras. Indeed, for most of human history, among those who knew little or nothing of medicine, rabies was merely another horror story in the same genre: a scream heard today in the next town over, quite possibly to resound in one's own town tomorrow.

In our more enchanted, pre-cinematic past, these types of stories spread not from the capacious minds and marketing budgets of Hollywood but out of tales told from house to house, town to town. These horrors were often related with the visceral sense (believed by both parties) that the menace in question was real and imminent. Stories evolved, too, as they spread, and so we can consider what remained after centuries of such "audience testing" as having a perverse sort of evolutionary fitness. It was not just the vampires and werewolves as such but a more generalized obsession with vicious half-human creatures, with dogs and wolves amok: girls (and boys) gone wild, familiar canids gone wrong.

The question, then, is not *who* the werewolves of the sixteenth century were, or the vampires of the eighteenth; the former were obviously victims of mass hysteria, the latter clearly corpses. The more relevant question is *why:* Why should it have been widely believed, and widely feared, that men were stalking the land as wolves? What is so terrifying about the vampire, a creature that, despite its human form, bites at the flesh of its victims? Why do dark forces so often manifest themselves in the shape of a dog? To such questions, our answer is the same as that of the good doctor Gómez-Alonso. The animal infection—the zoonotic idea—is mankind's original horror, and its etiology traces back inevitably to the rabies virus. Before our saga of the world's most diabolical virus careens into the nineteenth century, it is worth stopping for a moment to catalog the manifestations of this horror, from demon dogs to wolfish men and everything in between.

The original lycanthrope, from whose name the term derives, was Lycaon, the mythical first king of Arcadia. As the legend went, Zeus himself had descended to lodge in Lycaon's palace, and the king decided upon a wicked test of his guest's divinity. The king killed a boy and served him to Zeus at the table. On beholding the unappetizing cut he had been served, the god, immortally offended, slew fifty of Lycaon's sons with lightning bolts. Then, for good measure, he changed Lycaon into a wolf—a transformation that Ovid, in his *Metamorphoses*, describes in undeniably rabid terms:

> *Frightened, [Lycaon] runs off to the silent fields*
> *and howls aloud, attempting speech in vain;*
> *foam gathers at the corners of his mouth;*
> *he turns his lust for slaughter on the flocks,*
> *and mangles them, rejoicing still in blood.*
> *His garments now become a shaggy pelt;*
> *his arms turn into legs, and he, to wolf,*
> *while still retaining traces of the man:*
> *greyness the same, the same cruel visage,*
> *the same cold eyes and bestial appearance.*

Such an account conveys the Homeric *lyssa*, the infection with wolfish rage, except in this case the wolfishness is rendered quite literally. Likewise, many other ancient accounts of men becoming wolves, or of men possessed by animals, seem to stem from the inhuman ferocity with which some warriors were said to comport themselves in battle. Old Norse gives us the legend of the berserkers, ferocious fighters who wore the skins of bears or wolves atop their armor. Their rage was seen as a species of demonic possession, during which time they became immune to pain; one description of their prebattle mien has them foaming at the mouth, barking like wolves, chewing on the rims of their shields and sometimes gnawing them clean through. Similarly, centuries of Irish lore tell of the Laighne Faelaidh, a race of men who take the form of wolves whenever they please, killing cattle and devouring the flesh raw. A number of ancient Indo-European tribal names, such as the Luvians, the Lucanians, and the Hyrcanians, mean some variant of "wolf-men."

Some of the ancient accounts of wolf-men, and dog-men, shade into simple xenophobia. When Herodotus writes of the Neurians—a tribe in what is now eastern Europe, each member of which "changes himself, once in the year, into the form of a wolf," remaining thus for several days before changing back—it reads as the assertion less of a fearsome ferocity than of a subhuman curiosity. Another ancient chronicler, Ctesias of Cnidus, offers an account of a half-human tribe in India: "It is said that there live in these mountains dog-headed men; they wear clothes made from animal skins, and speak no language but bark like dogs and recognize one another by these sounds. . . . They couple with their women on all fours like dogs; to unite otherwise is a shameful thing for them." Strabo, a geographer from the first century B.C., wrote of the Cynamolgi, an Ethiopian tribe numbering some 120,000 dog-headed men who spoke in barks. Similarly, the Ch'i-tan, a tenth-century people in what is now Manchuria, believed that one of the regions to their north was "the Kingdom of Dogs," whose inhabitants "have the bodies of men and the heads of dogs. They have long hair, they have no clothes, they overcome wild beasts with their bare hands, their language is the barking of dogs." With some regularity did medieval maps place *cynocephali,* or "dog-headed men," in the edges of the known world, a practice carried out not just by Christian cartographers but also by their Muslim opposite numbers.

The easy explanation for such beliefs, and for the werewolf legend as well, is that these folk traditions employed the dog (and the wolf, her fierce or rabid cousin) as an expression of the so-called Other: that is, as a means by which to attribute a subhumanity to foreigners, outlaws, adherents to strange and scary creeds, and so on. And that explanation no doubt carries some truth. But isn't it telling that the animal chosen for "otherness" is, in fact, the opposite of strange? Indeed, what makes the demon dog such a powerful source of dread is precisely how familiar, in all senses of that word, the canine presence can be. When human beings keep dogs by choice, the dogs become our constant, often silent companions, living with us inside our strongholds. We become complacent about the animal nature that lurks in them still. But with the intrusion of rabies (or a passing squirrel, for that matter), such slavering essence can return with sinister immediacy.

Since dogs and humans possess an almost biological familiarity, having coevolved over millennia, even a strange and semi-wild dog today will take liberties with an unfamiliar human that other creatures will not. Barbara Allen Woods, a folklorist at the University of California in the late 1950s, built a taxonomy for the thousands of different European oral legends in which the devil appears in dog form. Regarding one such legend type, in which a demonic dog stalks a traveler, she observes:

> If there is any merit in the suggestion that legends of the devil in dog form are inspired by actual encounters with real dogs, it is most easily seen in stories about a night traveler who met with a demonic dog on the road one night. There is nothing extraordinary or mythical about such an incident. On the contrary it is entirely natural that a dog should be out trotting the deserted streets and paths.... Nor is it remarkable that a dog should follow a certain route; instead, it is typical of the canine species to make certain rounds. And it is perhaps least of all noteworthy that a stray dog encountered by chance should accompany a person for a time before jogging off on its own affairs. Yet, any or all of these normal characteristics can seem positively uncanny, especially when observed under eerie circumstances or in an anxious state of mind.

Ironically, the noted sixteenth-century demonologist Nicholas Remy turned this same reasoning on its head, in attempting to explain why evil spirits assume the form of dogs in the first place: "When [demons] go with anyone on his way, they most often take the form of a dog, which may follow him most closely without raising any suspicion of evil in the onlookers." Dogs have earned our trust, and we are used to their (sometimes unsolicited) companionship; what better vessel, in Remy's view, for a demon to exploit?

Woods's catalog is full of folktales in which a devil dog appears at moments of particular wickedness. A demon dog is encountered at a haunted place, such as a grave site, a churchyard, or a ruined castle. Or the appearance of the dog portends a death, even encourages someone to commit suicide. Humans shoot bullets at the demon dog, but it cannot be wounded. Dogs perch at the feet of cardsharps whose winnings flow from pacts with the devil. A dog lurks in front of a child's coffin and prevents his receiving a proper Christian burial.

Often the demon dog can be creepily communicative. A Danish boy in Frølund, when reading his parents' copy of a forbidden magic book, is interrupted by a noise in the hall. He opens the door to find a large black poodle, which gazes at the boy "with strange pleading eyes."* In one Swiss legend, two men see a dog watching a dance and

---

* Strange as this might seem to us today, a poodle appears frequently as the demon dog in old folktales. This association dates back at least to Goethe's *Faust,* which has Mephistopheles appearing to Faust in the form of a black poodle, which takes up residence with him and consistently interrupts whenever he tries to translate the Bible. When Faust tries to kick his new dog out, it reveals its true nature to him:

> In length and breadth how doth my poodle grow!
> . . .
> Huge as a hippopotamus,
> With fiery eye, terrific tooth!
> Ah! now I know thee, sure enough!

Freemasons, too, were thought to have sold their souls to the devil, who would attend their meetings in the form of a black poodle.

ask why he is there. The dog replies, matter-of-factly, that a fight is about to break out and someone will be killed; he, the devil, intends to claim that soul. In a similar Swedish tale, the dog is considerably more articulate. Two brothers from Sandåkra, after they commit perjury and escape detection, promise each other that whichever dies first shall return as a ghost, in order to tell the other what he has learned of the afterlife. Soon after the death of one brother, the second finds a large black dog sitting on the steps of his cottage. Knowing it is his brother, he asks the dog what he has found. "That which is once forsworn is eternally lost," replies the dog glumly. The living brother decides he must confess to his crime.

During witch trials, the accused often were found to have had canine "familiars" (that word again), demons who accompanied them in the form of dogs. Elizabeth Clarke, who during the seventeenth century admitted to having slept with the devil himself thrice weekly, was kept company during her sexploits by Jarmara, a white spaniel with spots, as well as by an ox-headed greyhound named Vinegar Tom. When the Devices—Alison, James, and Elizabeth—were convicted of witchcraft in 1612, all three of them claimed to have murderous dog familiars, with names like Dandy and Ball. In Alison's account of her dog's attack on a peddler, it is she who summons the dog to act but the dog who explains her options.

"What wouldst thou have me to do with yonder man?" the dog is alleged to have asked, as the peddler fled what he could tell would be an imminent attack.

"What canst thou do at him?" Alison replied.

"I can lame him."

"Lame him," replied the girl; and within forty yards the deed was done.

Notice the balancing act that is struck by this last tale, of the witch's canine accomplice. The dog must be possessed bodily by the most fearsome rage in order to carry out his bloodthirsty attacks, for example, to lame the peddler. And yet he must also be possessed spiritually of an almost human reason and capacity for understanding in order to present to the audience as properly and chillingly evil. It is the ancient dichotomy of the dog—between the intuitive, loyal companion and the savage, potentially rabid beast—with each pole of the dualism merely ratcheted out a notch. The uncanniness of the demon dog lies in his being simultaneously more familiar and more prone to insensate frenzy than the typical four-footed friend.

A similar formula undergirded the werewolf tales of the sixteenth century. Unlike the dog-headed men of maps, these were real people, often known to their alleged victims, who would testify with apparent sincerity that their neighbors had taken the form of vicious wolves. One oft-repeated tally, though perhaps apocryphal, puts the number of recorded cases in France at thirty thousand between 1520 and 1630. Regardless of the specific figure, history has bequeathed us enough specific cases to make clear that something like an epidemic was afoot. A sample:

1521. Two admitted werewolves, Pierre Burgot and Michel Verdun, stand trial in Poligny for many murders: of a four-year-old girl, of a woman gathering peas, and more still. Along with another lycanthrope confederate the two are convicted, burned.

1530. Near Poitiers, three enormous wolves set upon three young men, one of whom slices off a wolf ear in the melee. The following day, a known harlot in the town is observed to have lost an ear.

1541. A farmer in Pavia takes the form of a wolf and murders multiple victims. Upon his confession, the magistrates order the severing of his arms and legs, from which separations he dies.

1558. Near Apchon a huntsman, asked by a local gentleman to bring him some game, falls under attack by a wolf and severs its paw. Later, as he reaches into his bag to deliver this paw to his noble friend, he finds it has been transformed into a feminine hand—the hand, indeed, of the gentleman's own wife, who, when found to be missing it, confesses to being a werewolf. She is burned to ashes.

1573. The town of Dole, in the Franche-Comté region of western France, formally enjoins its peasantry to hunt down a marauding werewolf, authorizing the use of "pikes, halberts, arquebuses, and sticks."

1598. An entire family near Dole, the Gandillons, is executed for lycanthropy. The first to go, Pernette, had allegedly set upon two children, intending to devour them, but managed to slay only one of them, a four-year-old boy, with the pocketknife the child had brandished to defend his sister. Pernette is torn limb from limb by the citizenry.

Her crime draws the authorities' attention to her brother, Pierre, and to his son, Georges, both of whom confess (after what one suspects is rather insistent questioning) to having taken the form of wolves through the application of a salve. Pierre also has a daughter, Antoinette, who admits to starting hailstorms. All three are hanged, their bodies burned.

Meanwhile, two departments south, in the town of Châlons, a tailor is sentenced for having apparently lured, murdered, and eaten a numberless throng of small children. His alleged crimes are so terrible that the court orders the incineration of all the case records—and, naturally, of the tailor.

That same year, near Angers, a fifteen-year-old boy is murdered and a half-naked man, with long hair and beard, is taken into custody. This man, Jacques Roulet, admits to using a salve to transform himself into a wolf. He, too, is sentenced to death, though—in a sign the werewolf hunters of France have perhaps lost some of their moxie—the parliament in Paris later commutes his sentence to two years' incarceration.

1603. Jean Grenier, a teenager near Bordeaux, is arrested after terrorizing a series of local children, allegedly as both a boy and a wolf. Grenier's story was later recounted at length by Sabine Baring-Gould, a nineteenth-century English parson perhaps best known for composing the hymn "Onward, Christian Soldiers" but also the author of more than 130 books. Baring-Gould's *Book of Were-Wolves* (1865) to this day remains by far the most readable account of the werewolf phenomenon—so readable, in fact, that we hesitate to dwell upon the provenance of his elaborate narrative color and instead will simply draw upon it.

On a spring afternoon that year, as some young women are tending sheep ("the brightness of the sky," Baring-Gould writes, "the freshness of the air puffing up off the blue twinkling Bay of Biscay, the hum or song of the wind as it made rich music among the pines which stood like a green uplifted wave on the East . . . conspired to fill the peasant maidens with joy, and to make their voices rise in song and laughter, which rung merrily over the hills"), they encounter a redheaded boy of perhaps thirteen, perched on a log. Evidently poor, given his gaunt frame and tattered clothing, the boy nevertheless cuts a menacing figure, his prominent white teeth protruding from a grinning leer.

"I have killed dogs and drunk their blood," he tells the girls. "But little girls taste better; their flesh is tender and sweet, their blood rich and warm. I have eaten many a maiden, as I have been on my raids together with my nine companions. I am a were-wolf!" he goes on, as if that still needed spelling out. "Ah, ha! if the sun were to set I would soon fall on one of you and make a meal of you!"

The young women flee and tell others of this strange child they have encountered. As it happens, another local girl, Marguerite Poirier, knows the boy even better, having regularly tended sheep with him near their village of St. Antoine de Pizon. His name is Jean Grenier, she reports, and he has frequently terrified her with similar stories. Worse than that, he recently followed through on his threat to her: One day, when Jean was absent from his herding duties, a wolf attacked her and tore her clothes. The creature had red hair, like Jean's!

Grenier and his case are taken up by the parliament in Bordeaux, in an investigation that, as in other witch and werewolf trials of the era, yields a surprising array of confessions. A certain "black man" named M. de la Forest gave Grenier a salve and a wolf skin, he says, both of which he used to turn himself into a wolf. Besides his attack on Poirier, which he confirms in every particular, Grenier admits to having eaten three children, including an infant snatched from a cradle.

But as with the case of Jacques Roulet five years earlier, the parliament eschews execution, in favor of life imprisonment in a nearby monastery. Pierre de Lancre, a famous witch-hunter who had been involved with Grenier's trial, would visit the young man there in 1610. Grenier still copped to having once been a werewolf. Moreover, reported de Lancre, he "confessed to me also, in a straightforward manner, that he still wanted to eat the flesh of little children, and that he found the flesh of little girls particularly delicious. I asked him if he would eat it if he had not been prohibited from doing so, and he answered me frankly that yes he would." But the boy would never get his second helpings; soon after his interview with de Lancre, he would die in confinement, the cause unrecorded.

Richard Mead, one of England's most influential eighteenth-century physicians, published an account of rabies in 1702 that can only be described as lycanthropic. As with all fine horror tales, the case had been related to Mead secondhand, but (he assures us) by a man who was "very near of kin to the unhappy patient." In Scotland, the doctor recounts,

> a young man was bit by a mad dog, and married the same morning. He spent (as is usual) that whole day, till late in the night, in mirth, dancing and drinking: in the morning, he was found in bed raving mad; his bride (horrible spectacle!) dead by him; her belly torn open with his teeth, and her entrails twisted round his bloody hands.

The brevity of time between bite and neurological symptoms—less than a day!—dispels any notion that this was actually a case of rabies. The details of the attack, too, seem rather improbable. Rabies can elicit violence in human victims, to be sure, but these generally take the form of maddened outbursts, in which biting is uncommon. The concerted effort required to chomp open a human abdomen, not to mention dealing with the rush of fresh blood—it's all a bit more than the typical hydrophobic could handle.

Nevertheless, the parallels between this medical case report, on the one hand, and the then-popular reports of lycanthropy, on the other, are notable. Mead even goes so far, just a few pages later, as to cite the influence of the moon. "Looking over the histories of the many patients I have attended in this deplorable condition," he writes, "I observe about one half of the number to have been attacked with the spasms preceding the hydrophobia either upon the full moon, or the day before it." Like many physicians of his day, Mead attempted to apply to the human body the mechanical insights of Isaac Newton, whose mathematical demonstrations of the properties of physical objects had left a deep imprint on the late seventeenth-century psyche. Mead's theory was that the moon's gravity pulled the bodily fluids in various directions at various times, contributing to the patient's health or lack thereof. But despite this scientific (or at least quasi-scientific) framework, his nods to the moon in practice could seem arbitrary, even superstitious. Epileptic patients, he wrote, suffered spots on the face that resembled the dark patches on the surface of the moon; indeed, these spots "varied both in colour and magnitude, according to the time of the moon," and so would help the observant physician predict when seizures were imminent. Mead even cited approvingly a case, as related by an earlier author, of a woman whose beauty "depended upon the lunar force, insomuch that at full moon she was plump and very handsome."

The most striking aspect of Mead's lurid rabies case, though, is the setting of the scene: the wedding night, in which a young bride is deflowered in a horrifyingly unconventional manner. Domestic attacks, in which the assailed party is a spouse or lover, do sometimes figure in werewolf lore. One such tale is so widespread—having taken root from Transylvania to Uruguay—that folklorists gave it its own name: the "legend of the torn garment." In the most common version of this story a man, while riding home alongside his wife, unexpectedly hands the reins over to her and steps off into the bushes. The wife waits; suddenly a furious dog bolts out from the brush and bites down on her savagely. Afterward, alone, she makes her way home and finds her husband waiting there. As he walks to meet her with a smile, she spies scraps of her shredded dress in his teeth.

These sorts of intimate assaults, however, are considerably more common in vampire tales, where the dead spouse or lost love returns to haunt the living partner. Sabine Baring-Gould cites a vampire account from Baghdad, in the early fifteenth century, that bears more than a passing resemblance to the rabies tale of Richard Mead—though in this case it is the young woman who is driven to animal feastings. On the wedding night of one Abul-Hassan, the son of a wealthy merchant, the bride steals away from the marriage bed when she believes her new husband to be asleep. This she continues to do, night after night, until Abul-Hassan resolves to follow her. By moonlight he trails her to a cemetery, where he is faced with a terrifying tableau: a gang of ghoulish creatures, chowing down on corpses. With revulsion he sees his wife—who, Baring-Gould notes, "never touched supper at home"—playing "no inconsiderable part in the hideous banquet." The following night, Abul-Hassan confronts her with what he has witnessed. She lashes back quite literally with tooth and nail, tearing at his neck, attempting to drink his blood. At this, Abul-Hassan strikes and kills her; but three nights later, at midnight, she returns, again trying to sup at his neck. Only upon opening her tomb and burning her corpse is the vampire finally dispatched.

Before we move along to still more vampiric matters, it is worth reprinting the remedy for dog bite that Richard Mead advocated to forestall any onset of violent lunacy. First, the patient was to be bled from the arm, with nine or ten ounces removed. Second, a medicinal powder—a blend of black pepper and ground liverwort—was to be mixed into a half-pint of warm cow's milk and drunk by the patient each morning for four consecutive days. Finally, for a full month, the patient must bathe every morning in cold water. This last stage, Mead felt, was of the utmost importance, as demonstrated by the case of "a lusty young woman" treated by a certain Dr. Willis. Having been "raving mad seven or eight days," this woman, on Willis's orders, was "carried abroad at midnight, and thrown naked into a river: where she swam about without help for more than a quarter of an hour." Soon thereafter, reports Mead, she "recovered without the help of any other remedy." Presumably, this means the patient was no longer mad; whether she remained lusty, Mead does not say.

If the sixteenth-century werewolf epidemic had been a word-of-mouth hysteria, the vampire boom of the eighteenth century played out as a mass-media phenomenon. It was touched off by a series of published dispatches from eastern Europe, lands where vampirism served as a consistent force in the local folklore, written by Western correspondents, who reported on these strange happenings with horror. *Le Nouveau Mercure Galant,* a French newspaper, ran an account in 1694 of vampires "sucking the blood of people and cattle in great abundance." It went on: "They sucked through the mouth, the nose but mainly through the ears. They say that the vampires had a sort of hunger that made them chew even their shrouds in the grave."

Then, between 1710 and 1756, the great wave arrived: accounts from Prussia, Hungary, Silistra (in present-day Bulgaria), and Wallachia (in Romania; the haunt of Vlad the Impaler, whose name would later be appropriated by Bram Stoker for *Dracula*). Most famous among these accounts was the story of Arnod Paole, a dead Serbian soldier who locals believed had become a vampire. Due to the Peace of Passarowitz, signed by the Hapsburgs and the Ottoman Empire in 1718, the Serbian territory had been recently transferred to Austria, and so most of the Austrian soldiers detailed from the West were encountering Serbians and their lore for the very first time. Spurred by the local claims about Paole and others, an Austrian medical officer named Johannes Flückinger wrote up a brief report in 1732 called *Visum et repertum* (Seen and Discovered) that quickly saw wide dissemination and translation throughout western Europe. No doubt its appeal owed much to its persuasive form: a signed account by a soldier (and doctor, no less) who claimed to be laying out the facts soberly, just as he witnessed them. "After it had been reported that in the village of Medvegia the so-called vampires had killed some people by sucking their blood," Flückinger begins,

> I was, by high decree of a local Honorable Supreme Command, sent there to investigate the matter thoroughly, along with officers detailed for that purpose. . . . [The *haiduks* (that is, Serbian soldiers in the area)] unanimously recount that about five years ago a local *haiduk* by the name of Arnod Paole broke his neck in a fall from a hay wagon. This man had, during his lifetime, often revealed that, near Gossowa in Turkish Serbia, he had been troubled by a vampire, wherefore he had eaten from the earth of the vampire's grave and had smeared himself with the vampire's blood, in order to be free of the vexation he had suffered. In twenty or thirty days after his death some people complained that they were being bothered by this same Arnod Paole; and in fact four people were killed by him. In order to end this evil, they dug up this Arnod Paole forty days after his death—this on the advice of their Hadnack [or elder], who had been present at such events before; and they found that he was quite complete and undecayed, and that fresh blood had flowed from his eyes, nose, mouth, and ears; that the shirt, the covering, and the coffin were completely bloody; that the old nails on his hands and feet, along with the skin, had fallen off, and that new ones had grown; and since they saw from this that he was a true vampire, they drove a stake through his heart, according to their custom, whereby he gave an audible groan and bled copiously.

Intrigued by this account, Flückinger and his fellow officers accompanied the *haiduks* to the Medvegia graveyard and watched as they opened the graves of other suspected vampires, including the Hadnack's own wife, who had died just seven weeks before. Flückinger's team dissected a number of the corpses themselves, and it is clear from the report that they came away from this grisly work as believers. By the end, *Visum et repertum* has taken on the judgment of the locals, asserting that many of the corpses are in a "condition of vampirism."

Most vampire reports of the era are essentially similar: in each the shocking observation, made by a dispassionate Western observer, is that the vampire's corpse looks surprisingly intact, with fresh blood lingering around its mouth. But the American scholar Paul Barber, in his wonderful 1988 book, *Vampires, Burial, and Death* (from which the above excerpt of Flückinger's report is drawn), makes a very compelling case that these reports, even those by medical men, simply misapprehend the ways that bodies can decompose after death. The epidermis often slips off, revealing the dermis, which resembles a "second skin"; the nail beds resemble new nails. And, most important, bodies are prone to swell, pushing what looks like fresh blood—in fact reliquefied blood—out from the nose and mouth. The "chew[ing] of shrouds in the grave," as *Le Nouveau Mercure Galant* put it, is in fact the sound of swelling bodies gurgling and bursting; the ever-present "groan" of a staked vampire is no more and no less, in Barber's view, than the release of pent-up gases.

Barber's other key point is that many of the attributes we associate with vampires—indeed, quite a few of those cited by Dr. Gómez-Alonso—are in fact creations of the fictional vampire, as drawn by Western writers of the nineteenth century. It's true that Eastern folklore did sometimes assert that the vampire changed shape into animals, but not always, and not generally as dogs: a tale from Siret, in northern Romania, has the vampire becoming a cat in order to escape detection, while another folklorist lists the animal forms of the vampire as "wolf, horse, donkey, goat, dog, cat, pullet, frog, butterfly." The snarling black dog form of the vampire, and the yelping dogs that greet the vampire's presence as he silently treads down gloamy country paths, are fictive creations of a particularly English character.

More crucially for our purposes, the vampire's bite—so key to our understanding of him today—is largely absent from folkloric accounts of vampires. If he bites at all, it is at the chest or torso; more often, he smothers his victims to death or attempts to smother them as they sleep. Indeed, sleep, that nightly dress rehearsal for death, is so often the meeting place between vampire and victim. In his original incarnation, it seems to have been the unbridgeable crossing of death, rather than the uncanny animal furies of man, that conjured up the vampire in the popular imagination. But during his westward migration, and his rebirth during the nineteenth century as a fictional force, the vampire changed into something new and yet, for the purposes of our tale, more biting.

The vampire we know today was born, interestingly enough, at the same time and in the same place as his famous gothic friend, Frankenstein's monster. It was the summer of 1816, in Switzerland, when five bons vivants from England—the poets Lord Byron and Percy Shelley; Shelley's paramour, Mary Godwin (soon to be Mary Shelley), and her stepsister, Claire Clairmont, then pregnant (though this was not known at the time, perhaps not even to herself) with Byron's child; and Byron's personal physician, John Polidori—all gathered at the Villa Diodati, a manor house near Lake Geneva that Byron had rented for the summer. It is hard to see what peace any of them might have hoped to find at this lakeside retreat, given the interpersonal drama that stalked Byron no matter how far he tried to flee it. Despite his attempts at abandoning Clairmont in England, it was she who convinced the Shelleys they should visit him, and so Byron was forced to deploy Polidori as a human shield, to keep Clairmont from cornering him alone. Meanwhile, he and Polidori were at war with each other, even though the doctor was nominally in his employ: after Byron laughed at the doctor's spraining his ankle, Polidori retaliated by clocking Byron with an oar. To make matters worse, the outside world believed Byron's checkered personal life to be even more dramatic than it actually was. The proprietor of a hotel across the lake actually rented out telescopes for the purpose of spying on the famous writer's carnal depredations. When some tablecloths were hung to dry on the villa's balcony, the hotel gawkers reported them as ladies' petticoats, which they naturally assumed were shed on arrival to the villa as the price of admission.

One night, four of the five companions—Clairmont excepted—resolved to engage in a writing project. They had all been reading stories aloud to one another from a French book of supernatural tales called *Fantasmagoriana,* and Byron evidently felt that the assembled company could do better. "We will each write a ghost story," he suggested; and, as Godwin recounted later in her introduction to *Frankenstein,* "his proposition was acceded to." Her own contribution, of course, soon grew into her most famous creation. Shelley's and Byron's notions seemed halfhearted, relative to their creative powers, and neither poet chose later to expand upon his. Then there was "poor Polidori," as Godwin patronizingly called him: he "had some terrible idea about a skull-headed lady, who was so punished for peeping through a key-hole—what to see I forget—something very shocking and wrong of course." Her account leaves us to believe that her ghost story alone, from the four conceived during those days at the Villa Diodati, survived to haunt the reading public.

In fact, though, later that summer, one of the dead ideas found new life, albeit in the hands of another of the participants. Having tucked his own "terrible idea" away for future use (it would play a minor role in a larger novel, though still to little acclaim), Dr. Polidori found himself ruminating on Byron's fragment. It was a very simple ghost story, hardly developed at all. Two friends from England travel to Greece, and one dies while there. Before his demise, the dying man asks his friend to swear never to reveal at home that he is dead, and the friend agrees. But back in England, he soon sees his dead friend return to his place in society, and so the living man is cast into agony: he can never tell his friends—even his own sister, who begins to fall in love with the undead man—that they are trafficking with a specter.

Goaded by a lover, Polidori spent a few days embellishing this bare outline into a ghost story both more frightening and more cutting than originally conceived. It was more barbed, in that Polidori clearly modeled his undead villain on his employer (soon to be ex-employer, for the two quarreled constantly), Byron himself. The dying friend became Lord Ruthven, a dissolute and financially troubled aristocrat, a caddish seducer of women. The name Ruthven itself was a dead giveaway: it was the same name given to the Byron figure in *Glenarvon,* a thinly fictionalized tell-all about the poet that was penned by Lady Caroline Lamb, one of his many recent entanglements, and published right before Polidori was writing.

But the retooled tale was more terrifying, too, because Byron's specter became, in Polidori's hands, not merely a ghost but a vampire, a figure then known well in popular intrigue but not yet so well in fiction. Polidori's fictional vampire, though not the first in English, would become the template for essentially all the vampire fiction (and, later, film) that was to follow. In his brio to satirize Byron, Polidori was led to make two brilliant metaphorical connections that persist to this day. The first is the vampire as aristocrat, as a man whose dealings with the rabble are confined to predations upon their very flesh. The second, and more crucial, is the vampire as seducer: a man whose attitude toward women is driven by unslakable, quasi-sexual (or literally sexual) appetites. And yet in the moment of consummation, as it were, the vampire takes on a *lyssa*-like rage, as the innocent male protagonist discovers:

> He was lifted from his feet and hurled with enormous force against the ground:—his enemy threw himself upon him, and kneeling upon his breast, had placed his hands upon his throat when the glare of many torches penetrating through the hole that gave light in the day, disturbed him.

Soon thereafter, the female victim is discovered: "There was no colour upon her cheek, not even upon her lip; yet there was a stillness about her face that seemed almost as attaching as the life that once dwelt there:—upon her neck and breast was blood, and upon her throat were the marks of teeth having opened the vein."

These two attributes, nobility and lust, also define our own vampires, from Bram Stoker to Anne Rice to Stephenie Meyer. Vampirism is a dark, animal undercurrent that haunts the human—even the most refined among us, and even in our closest relationships. The vampire's bite shocks most for its shattering of our admiration, our domesticity, our intimacy. Beyond the profusion of vampire fiction that spun out of Polidori's tale (published as *The Vampyre*) and reached its apex with *Dracula,* the vampire also came to function as a powerful trope in less fantastical writing as well. Much like with the *lyssa* of Homer, or the *rage* of *The Song of Roland,* the vampire stalks through nineteenth-century English literature as a ready-made metaphor for the animalistic force undergirding the passions of men—or, as the case may be, of women.

It would not be until the very end of the nineteenth century that rabies' most ancient host—the bat—would find a permanent home in vampire tales. The association had been made for centuries, though, by those who followed news from the Spanish New World. An early sixteenth-century account of Hispaniola, penned by the historian Gonzalo Fernández de Oviedo y Valdés and published in abridged form during the 1520s, described such curiosities as the pineapple, the hammock, and tobacco. His account also introduced Europeans to a terrifying variety of bloodsucking bat. "Usually they bite at night," Oviedo reported, "and most commonly they bite the tip of the nose or the tip of the fingers and toes, and suck such a great amount of blood from the wound that it is difficult to believe unless one has observed it. . . . The wound itself is small, for the bat takes out only a small circle of flesh." Translations of Oviedo's abridged history found popularity throughout western Europe during the 1550s; over time, the Spanish conquistadores would come to call the bats *vampiros*, because of their resemblance to Europe's mythic monsters.

Comprising three species confined to the tropical and subtropical regions of the Americas, vampire bats are unique among mammals for their habit of subsisting on the blood of other warm-blooded vertebrates. Oviedo's description of their feeding behaviors is impressively accurate. The bats do preferentially bite the capillary-rich tips of fingers, toes, and noses; and through a small circular aperture made in the victim's skin, they indeed can lap large quantities of blood for their size—thanks to an anticoagulant in their saliva that also can lead to excessive bleeding in their victims after they drink their fill and flap away. (When this anticoagulant was discovered in the twentieth century, it was puckishly named draculin, a moniker that has stuck.) Almost certainly these bats harbored rabies at the time of the Spanish conquest, and Oviedo's account provides some support of that fact: he calls their bites "poisonous" and reports that "some Christians died" from the poison before the natives explained their local cure, namely cauterization.

By the early nineteenth century, tales of vampire bats circulated widely in the English-speaking world. J. G. Stedman, in a 1796 account of his years in Suriname, describes his encounter with a blood-feeding bat in fantastical terms. "On waking about four o'clock this morning in my hammock," he writes,

> I was extremely alarmed at finding myself weltering in congealed blood, and without feeling any pain whatever.... I had been bitten by the *vampire,* or *spectre,* of Guana, which is also called the *flying dog* of New Spain, and, by the Spaniards, *perro-volador.* This is no other than a bat of a monstrous size, that sucks the blood from men and cattle when they are fast asleep, even, sometimes, until they die; and as the manner in which they proceed is truly wonderful, I shall endeavor to give a distinct

> account of it.—Knowing by instinct that the person they intend to attack is in a sound slumber, they generally alight near the feet; where, while the creature continues fanning its enormous wings, which keeps one cool, he bites a piece out of the tip of the great toe, so very small, indeed, that the head of a pin could scarcely be received into the wound, which is, consequently, not painful; yet through this orifice he continues to suck the blood, until he is obliged to disgorge. He then begins again, and thus continues sucking and disgorging till he is scarcely able to fly, and the sufferer has often been known to sleep from time into eternity. . . . Having applied tobacco-ashes as the best remedy, and washed the gore from myself and from my hammock, I observed several small heaps of congealed blood, all around the place where I had lain, upon the ground; upon examining which, the surgeon judged that I might have lost at least twelve or fourteen ounces during the night.

Around the same time, the Spanish painter Francisco Goya was using spectral, bat-like figures to symbolize vampiric forces. Great shadowy bats hover above a slumped figure of Reason, in *The Sleep of Reason Produces Monsters,* and also, in *There Is Plenty to Suck,* behind three murderous hags as they prepare to consume a basketful of babies. His *Los caprichos* illustrates a series of vampire-like figures in the act of devouring sleeping innocents. In 1804, William Blake depicted a vampire bat in two engravings accompanying his poem *Jerusalem* to symbolize what he calls the Spectre—the divisive and annihilating energies that cannibalize the human psyche. But it took scientists until 1810 to provide a description of the hematophagous (that is, bloodsucking) bat. Even in 1839, when Charles Darwin commented on the feeding habits of a *Desmodus* bat during his travels aboard the *Beagle,* he noted that the "whole circumstance has lately been doubted in England; I was therefore fortunate in being present when one . . . was actually caught on a horse's back."

It was left to the British vampire novel to formalize the relationship between the undead creature and its Latin American namesake. The eventual cover of James Malcolm Rymer's *Varney the Vampyre,* which began as a horror serial that ran between 1845 and 1847, sported four Satan-headed bats, hovering menacingly around the skeletally dapper Sir Varney as he stands poised to sup at the throat of the raven-haired beauty drowsing beneath him. And then Bram Stoker's *Dracula,* in 1897, made the connection unequivocal, with the count's presence often signaled by his bat rather than by his human form. Ever since, the fictionalized vampire has traveled in the abundant company of bats, whether in novels, in Hollywood, or on *Sesame Street*—though in that last instance, the number of bats can always be readily counted.

A New York City policeman shoots a rabid dog on Broadway. From *Harper's Weekly*, 1879.

4

CANICIDE

In 1847, the American evangelist Alexander Campbell traveled through Europe, sending home sporadic reports to be published in the monthly magazine he founded, entitled the *Millennial Harbinger*. In the northern wilds of Scotland, Campbell came across a melancholy scene, which he painted with a sensitivity and delicacy befitting a man of the cloth. On his long journey from Aberdeen to Banff, he stopped to visit the lovingly cultivated but nearly uninhabited estate of James Duff, fourth Earl of Fife, who lived there at the musty old age of seventy-one without family and with few servants. A nocturnal creature in this enormous, unfinished castle, Duff typically awoke at five in the afternoon and returned to his bed by five in the morning. "One cannot conceive," Campbell wrote, "why he should live in the midst of such fine gardens and groves, ornamented with beautiful walks, summer-houses, alcoves; bowers, jetteaus, &c., as environ his splendid residence, to be surveyed by himself for an hour or two in the evening of the day." But, the pastor allowed, Duff did have one reasonable excuse. Forty-two years beforehand, his wife—the former Maria Caroline Manners, a legendary beauty—had died at the age of thirty, just six years after their marriage, leaving him no children. The cause was "the canine madness"; she was "bitten by her own rabid lap-dog."

One can hardly blame Duff for failing to recover or remarry. His wife's death in 1805 had left Edinburgh startled and not a small measure scandalized, as much for its peculiar beginnings as for its horrific conclusion. At some point during the previous year, four dogs of the Duff family had suffered bites from a rabid attacker. Three of these dogs belonged to the earl and were immediately put down. But the lady could not bear to lose her own lapdog, a French poodle named Pompey—a fitting allusion, it would seem, to unforeseen disaster. The creature was spared, but it proved to be a tragic pardon.

Many months later, the fateful symptoms began to appear in the lapdog and then, sometime later, in the lap's owner. Rumor had it that to end her agonizing spasms, it was necessary to smother the splendid lady to death, though her doctors later insisted this had not been the case. Known to all as a dazzling beauty, Mrs. Duff was subsequently memorialized in a popular engraving, well trafficked for decades afterward, that showed the young lady stepping atop the crest of a globe, bearing aloft a sash and attended by cherubs whose presence connoted an ascent (as one admirer noted years after her death) to "another, and a better world."

The precise mode of transmission remains disputed to this day. Soon after the event, the general opinion was that Pompey had nipped her on the tip of the nose. One Edinburgh wag, Charles Kirkpatrick Sharpe, quipped that "no nose was so much talked of since the days of Tristram's Don Diego" and went on to describe the town's general uproar:

> Not a grain of *rouge* was left on a single cheek in E[dinburgh] with weeping; not one female tongue ceased talking of the catastrophe for a week. "Oh, she was such a sweet creature!" She had bought a whole cargo of silk stockings the day before she fell ill, and expected new liveries for her footmen every moment.

> Indeed, she had not one fault on the face of the earth. She was to have been at a ball the very night she died.

But later there seems to have been a consensus, developed among doctors, that the poodle was innocent of even a bite and that its friendly licks alone had spread the dreaded disease. An 1830 paper in the *Lancet* laid out this particular theory in detail. "She had a small pimple on her chin, of which she had rubbed off the top," wrote William Lawrence, one of the journal's founders; "and allowing the dog to indulge in its usual caresses, it licked this pimple, of which the surface was exposed." Subsequent reports (perhaps following Lawrence) also specifically cited a pimple as having been the aperture to infection.

True or false, this invocation of unseemly acne on the fair Mrs. Duff's visage also comported with the general disapproval, among nineteenth-century medical men, of the intimate canine congress to which lapdoggery led. The act of allowing a cur to lick a human face was, to their minds, the height of uncleanliness. "Not only a most disgusting, but a dangerous practice," intoned one medical author, in discussing the Duff case; another called it "degrading" and "reprehensible" and went so far as to say, "I unequivocally condemn an indiscriminate attachment to, or imprudent fondling of dogs."

In their scorn, these men of science were responding to a genuine sea change in the way pets, and in particular dogs, were regarded in the industrializing precincts of Europe and the United States. With the rise of a middle class, no longer tied to farms and their necessarily instrumental approach to animals, the treasured, pampered pet no longer figured as merely a luxury of the upper classes. In 1840s Paris, whose human population was just shy of a million, there were believed to be some 100,000 pet dogs. The nineteenth century saw the rise of dog shows (with more than three hundred held each year in England by century's end) and "dog fancy" in general, a practice that could be joined even by those of modest means. Ever more citizens during the sweep of this century found themselves inclined to agree with Lord

Byron's epitaph for his beloved Newfoundland, Boatswain: "To mark a friend's remains these stones arise; I never met but one—and here he lies." (Boatswain expired from rabies in 1808, at which time Byron had him interred on the grounds of Newstead Abbey, the family's ancestral home; three years later, Byron specified in his will that he be buried at Boatswain's side, though this instruction was later countermanded.)

This transformation in pet-keeping was taking place at the same time that Europeans, with the increase in literacy and the explosion of the press, began to learn more about life in their ever-expanding colonies. Perhaps inevitably, the distinction between the familiar, domesticated pet and the ungovernable wild animal came to be seen as analogous to that between civilized and savage races. When Charles Darwin, in his 1868 treatise on domestication, remarked upon the tendency of half-bred animals to revert to a wild nature, he compared this to "the degraded state and savage disposition of crossed races of man" and approvingly quoted an assessment made to the famed Dr. Livingstone by a Zambezi native in Africa: "God made white men, and God made black men, but the Devil made half-castes." With animals as with man, domestication and fine breeding were seen to bestow a moral as well as a physical fitness.

By insinuating itself into domestic tranquillity—by effecting the fall of Pompey, as it were—rabies presented itself as a shocking subversion of this order. As such, it became an object of disproportionate panic throughout the nineteenth century. Reports of allegedly mad dogs studded the newspapers. Stories of actual hydrophobic expiry were granted full columns stuffed with florid detail. Neighborhood councils formed to beat back the scourge of feral dogs, even as tuberculosis and cholera cut wide and far more fatal swaths through their same streets. As the historian Harriet Ritvo has pointed out, a person of that era (in England, at least) was ten times more likely to die of even murder than of rabies. But a death at the hand of man seemed far less horrible to contemplate than one suffered in the jaws of the devil.

Compounding the terror was the fact that science understood rabies little better at the start of the nineteenth century than it did at the end of the second. Good old Soranus of Ephesus, as handed down to us by Caelius Aurelianus, had a more sensible take on the causes and nature of hydrophobia than did many medical men of 1800. When Benjamin Rush, one of the United States' most esteemed doctors and medical authors of revolutionary times, published his thoughts on the disease near the turn of the century, he began with a list of twenty-one supposed causes of hydrophobia, and the bite of a rabid animal did, mercifully, place first. But the balance of the list included "cold night air," "eating beech nuts," "a fall," and "an involuntary association of ideas." Seventeen centuries after Soranus and Suśruta had each succeeded in largely differentiating rabies from other maladies—not fully, to be sure, and not quite with rigor, but nevertheless judiciously—medicine was once again having difficulty in doing so.

Though a colonist, Rush was far from a rube by European medical standards. For his doctoral training he braved the ocean voyage to Edinburgh, whose university housed one of the late eighteenth century's finest medical schools. There he was steeped in the teachings of the top European minds of the day, in particular Herman Boerhaave, the revered Dutch theorist who had died in 1738 but remained the towering figure in medical instruction throughout Europe. The chair of Edinburgh's medical school was William Cullen, a world-renowned doctor and thinker in his own right, though his views diverged only slightly from Boerhaave's. (On those instances when they did diverge, Cullen later recalled, he was immediately decried as "a whimsical innovator" whose apostasy would "hurt myself and the University also.") Still influenced by Newton and his revolution in physics, these eighteenth-century doctors saw the body mechanistically, as a sort of hydraulic contraption of solids interacting with fluids, or "humors." In their view, diseases often were corruptions of these humors, which could become overly acidic or alkaline. In assigning causes, the theory looked disproportionately to diet, reasoning that just as food could undergo insalubrious transformations outside the body, through spoilage and so on, it could also effect corruption on the inside.

From November to May, Rush imbibed these theories six days a week, with almost no leisure time to speak of. The typical day involved studying all morning, then classes and hospital rounds in the afternoon, followed by still more reading until midnight. Edinburgh also schooled Rush in the experimental method. His chemistry teacher, Joseph Black, himself had used the occasion of his doctoral dissertation to prove the existence of carbon dioxide (or "fixed air," as he called it), a result that opened the door to the discovery of oxygen. By comparison, Rush's own dissertation, an inquiry into the acidity of the stomach during digestion, was rather less impressive in both methodology and results. "Having dined on beef, peas, and bread," he wrote, "I puked up, about three hours afterwards, the contents of my stomach, by means of a grain of tartar emetic, and found them not only acid to the taste, but likewise that they afforded a red color, upon being mixed with the syrup of violets—an invariable mark this, of acidity, among chemists!"

Nevertheless, when Rush returned to America, he was appointed professor of chemistry by the College of Philadelphia. This made him, at twenty-three, the first chemistry professor in what was soon to become the United States. A staunch supporter of American nationhood, Rush eventually had the opportunity, as a member of Pennsylvania's congressional delegation of 1776, to sign the Declaration of Independence, and during that tumultuous era he increasingly wedded his politics to his profession. After the Tea Act, Rush played upon his status as a noted physician to showily condemn tea itself, as having "mischievous effects on the nervous system," and as the Revolution loomed, he applied his chemistry, too, penning three essays about how to make saltpeter—the crucial ingredient in gunpowder—from dried tobacco stalks; Congress later distributed these essays as a pamphlet, with an introduction by Ben Franklin. During the Second Continental Congress, Rush inoculated Patrick Henry against smallpox, and at

Trenton, after George Washington's famous crossing of the Delaware, Rush was on hand to perform battlefield medicine.*

Rush believed that rabies was, in essence, a "malignant state of fever." In this opinion he was following a tradition, beginning at least with Boerhaave, that fever could be the immediate (though not always the root) cause of hydrophobia. Boerhaave attributed this fever to "inflammation," a catchall concept in his mechanistic theory. Cullen, following this notion, developed the idea that such fevers were caused by "overstimulation" due to excess blood. The natural prescription, therefore, was bleeding, which is indeed what Rush advocates for the rabies sufferer. Rush's main evidence for hydrophobia being an "inflammatory fever" is his observations of the blood, which, when drawn, exhibits both "size" (a term then for viscosity) and yellowness in its serum. He recalls that the blood "was uncommonly sizy in a boy of Mr. George Oakley whom I saw, and bled for the first time, on the fourth day of his disease, in the beginning of the year 1797. His pulse imparted to the fingers the same kind of quick and tense stroke which is common in an acute inflammatory fever."

One of Rush's medical students, James Mease, held a somewhat more accurate view—that hydrophobia was a disease of the nervous system—and, awkwardly for the two men, this view had seemingly been endorsed by Rush before the teacher later contradicted his pupil. Mease had laid out this theory in his 1792 doctoral thesis on the disease, which he had dedicated to Rush and for which Rush had even supplied a preface.† In 1801, chagrined to see that his old professor had now

---

* Later, after a long feud with William Shippen, the army's top medical man, Rush would sour on Washington's leadership of the army; when he expressed these sentiments to Patrick Henry, then governor of Virginia, Rush's name became linked with the so-called Conway Cabal, which aimed to replace Washington with another general, Horatio Gates. Rush's reputation as a patriot has unfairly suffered as a result.

† The preface does, it should be noted, read suspiciously like boilerplate. "I cannot consent to the publication of your ingenious dissertation," it begins, "without requesting you to allow me room enough in your preface, to express the great pleasure I derived

published a contrary opinion, Mease furiously produced a second pamphlet on rabies, this one explicitly addressed to Rush. Point by strenuous point he set out to rebut his former mentor. The woundedness of Mease's tone as he did so (beginning with his florid prefatory note to Rush himself, telling the great physician that he has mistakenly returned to principles "previously destroyed by yourself") was understandable given the personal connection between the two men. A family friend, Rush had known the young Mease since infancy and personally treated him through several serious childhood illnesses. "One of the first things I can remember," Mease later recalled, was Rush "calling me 'his boy,' and he used frequently to say that I should be his apprentice." This latter prediction, which Mease averred was "probably in jest," was eventually realized when Mease enrolled for medical training at the College of Philadelphia.

In many respects, Mease's gloss on rabies is hardly more helpful than Rush's. He asserts that the illness can sometimes generate spontaneously in dogs, a widespread theory (as we will discuss later in this chapter) that seriously set back attempts to contain and control the disease. And by contemporary standards, his preferred treatment for hydrophobia—powdered jimsonweed—is every bit as daft as bleeding. But Mease does draw one very perceptive comparison that had eluded medical authors until then. Noting the irregularity of the onset of hydrophobia (the fact that the time between bite and symptoms is usually weeks, or as long as months) Mease realized that another illness involving puncture wounds behaves in a similarly odd way: namely, tetanus, or lockjaw. From what we know today, the two are different in many crucial respects: rabies is a virus, which travels up the nerve sheathings, whereas tetanus is a bacterium, with the chemical poison released by that bacterium, rather than the bacterium itself, journeying up the nerves. But rabies and tetanus are cousins, as it were, in that

from reading it. It will be resorted to hereafter as a repository of facts and opinions upon the disease of which it treats."

they are among the few pathogens whose malign effects spread along the nerves instead of the bloodstream, creating the strange phenomenon of a disease whose time of onset depends on the distance of the wound from the head.

Such minor flashes of insight aside, all these millennia of medical theorizing about rabies had yielded little usable knowledge. The reason for this, of course, is that medicine had yet to embrace the most crucial insight of all: the very existence of viruses and bacteria, agents of disease unseeable by the naked eye. At the time of Rush and Mease, science possessed neither the complex optics nor the practical wisdom required to make this leap. Less than a century later, seventeen centuries of stasis in our understanding of rabies would be obliterated—and largely by the dogged efforts and remarkable genius of a single man.

During one particularly bad rabies outbreak in England, the *Liverpool Daily Post* published a comical ditty, called "The Two Dog Shows," that captured perfectly the stark dichotomy in the canine universe, between the increasingly cherished domestic and the increasingly terrifying feral. The poem began:

> *All London for the past few days, as you, of course, well know,*
> *Has crowded Islington to see the Great Dog Show;*
> *Which has been totally eclipsed by Liverpool: we find*
> *They've had a dog show there for weeks of quite another kind,*
> *One which is "open every day" in all the streets and lanes;*
> *And which consist of tortured dogs, and dogs without their brains.*

The nineteenth century was indeed the best and the worst of times for the Western dog, the most fortunate of which had finally been invited onto the foot of the bed, but the least fortunate of which met brutal ends in large numbers at the hands of exercised humans. In France, whose reputation as a locus of pet pampering was well deserved (one contemporary book on dogs stated, on its first page, "*Le chien est une machine à aimer,*" that is, "The dog is a love machine"), rabies fears in Paris led to "Great Dog Massacres," sometimes called "canicides"—the tally in 1879 alone was 9,479 killed. In England, the preferred method of dispatch was beating the dogs to death with truncheons. Neil Pemberton and Michael Worboys, two University of Manchester historians who have studied the history of rabies in Britain, note that from the perspective of dog lovers, terror of the disease had "turned ordinary people into murderers."

With little surety to be found in medicine, observers were left to formulate their own theories about which dogs should and should not be trusted. To some, domestication itself was seen as the enemy—"Constantinople and Africa are rabies-free," it was often noted, erroneously—even as reports were coming in from India that the disease ran rampant among wild dogs and jackals. Some people believed that the best-kept dogs were the most dangerous—that too much idleness, combined with overfeeding, predisposed them to the disease; that the inbreeding of purebred dogs "exhausted" their "nervous system" and made them susceptible. (As one correspondent remarked in the pages of what must be the best-titled publication in history, the *Annals of Sporting and Fancy Gazette,* "Hydrophobia makes its appearance . . . in dogs which exist in a state of confinement which are kept in towns and take little exercise.") Different theories fingered different breeds as the primary culprits: retrievers, in one view; foxhounds, in another.

Most common of all, of course, was to blame the dogs of the lower classes. This was largely the view of animal-welfare advocates, who naturally believed that mistreated dogs, such as those made to pull carts (still a surprisingly common practice) or fight in rings, were the most susceptible. But in most people's minds, it seemed clear that the degradation of the dogs was merely analogous to the degradation of their owners. One letter writer to the London *Times* remarked that these curs "infest our streets unmolested, creating noise, filth, and general annoyance; nor is this all, the peculiar sexual intercourse of the species renders them very dangerous. I have seen a whole gang of curs so strong excited, as to be little short of mad—so furious, indeed, as to change their very natures." Wrote another correspondent to the *Times,* "If these no-breed curs, which are at least two hundred to one, were destroyed, there would be little fear of hydrophobia." In the 1850s, France created its dog tax for the stated purpose of discouraging dog ownership among the poor. Britain had a similar tax, though it was levied only on dogs more than six months old, which wound up worsening the problem it aimed to fix: families would often acquire adorable puppies and then dump them on the streets, leaving them to grow up wild.

In the end, no dog could be trusted. Indeed, some experts cautioned that uncommon affection, of all things, could herald the onset of rabies. George Fleming, the British veterinarian whose 1872 treatise on hydrophobia was probably the definitive English-language survey in the pre-Pasteur era, warned against the "Judas' kiss" of the rabid pet, explaining that

> its instinct impels it, at times, to draw near to its master, as if to ask for relief from its sufferings; and, if permitted, it willingly tenders its recognition of the care bestowed on it by licking the hands or face. But these are perfidious caresses, against which every one should be warned.

The French physician G. E. Fredet took this admonition further. In his view, this behavior in the early stages of rabies was not occasional and not confined to masters alone. Instead, such dogs "invariably express an exaggerated attachment and devotion to everyone who approaches them." That is: even the friendliest dog on the street, or in an acquaintance's home, might suddenly deliver a bite (or lick!) that became a death sentence.

Given this dual nature of the dog, it is perhaps easy to imagine why fiction of the gothic persuasion, when hoping to conjure an atmosphere of gloom, would trot out so many snarling curs. Few gothic novels play this card better, or at least more often, than Emily Brontë's *Wuthering Heights*. In the very first pages, when the tenant Lockwood walks up to visit his brooding landlord, Heathcliff, the visitor's unwelcomeness is underscored by the hostility of his host's dogs—in particular the mother dog, a "liver-coloured bitch pointer" that "broke into a fury, and leapt on my knees" and that (along with half a dozen other "four-footed fiends, of various sizes") Lockwood is left to fend off with a fireplace poker. During the ensuing snowstorm, when he tries to escape Wuthering Heights by borrowing one of the house's lanterns, the servant sics the slavering dogs on him to prevent the theft; on order, "two hairy monsters flew at my throat, bearing me down, and extinguishing the light."

It continues this way through much of the novel. A similar dog attack, we soon learn, played a pivotal role in the tale of Heathcliff and his lost love, Catherine Earnshaw. Although the two were inseparable for many years as children, Catherine's life takes a turn when Skulker, the family bulldog of the wealthy Lintons, sets upon her savagely outside the Linton home. Heathcliff attempts to free her, prying at the dog's jaws with a rock, but the beast holds fast, "his huge, purple tongue hanging half a foot out of his mouth, and his pendant lips streaming with bloody slaver." Found mangled in the grip of this beast, Catherine is briefly adopted by the Lintons, who nurse her for five weeks—during which time she takes up, at least in part, not only their upper-class values but also an attachment for their son, Edgar, who thereafter will rival Heathcliff for her love.

And indeed, during the final days of Catherine's life, in the throes of a childbirth that will kill her, Heathcliff is possessed by what Brontë describes as something akin to canine madness, as the servant Nelly recounts:

> Catherine made a spring, and he caught her, and they were locked in an embrace from which I thought my mistress would never be released alive. In fact, to my eyes, she seemed directly insensible. He flung himself into the nearest seat, and on my approaching hurriedly to ascertain if she had fainted, he gnashed at me, and foamed like a mad dog, and gathered her to him with greedy jealousy. I did not feel as if I were in the company of a creature of my own species; it appeared that he would not understand, though I spoke to him; so, I stood off, and held my tongue, in great perplexity.

Emily Brontë herself beheld both aspects of the nineteenth-century dog while growing up in her father's parsonage. On the one side, there was her bulldog, Keeper, whose penchant for napping on the family's beds upstairs was not tolerated but nevertheless fondly recollected. On the other side, there was the strange dog, clearly in distress, perhaps thirsty, to whom Emily offered water one day as a child. The dog, rabid, bit her. Emily immediately strode into the kitchen, grabbed an iron that Tabby, the family cook, kept heated there, and cauterized her own wound. She told no one of the incident until much later, after the danger of infection had presumably passed.

After Emily's death, her sister Charlotte brought the incident into her novel *Shirley*. The bitten party became not a child but a young woman, the novel's fierce and wealthy protagonist, Shirley Keeldar; the bite, and her concealment of it, becomes the pretext by which the headstrong Shirley softens her heart to marrying the penniless tutor Louis Moore, who has proposed to her. She fears that she will die, and she confesses this to Louis. As he comforts her with what today seems laughably false confidence ("I doubt whether the smallest particle of virus mingled with your blood: and if it did, let me assure you that—young, healthy, faultlessly sound as you are—no harm will ensue"), it is clear that Shirley's resistance to his romantic advances has melted. Charlotte told one of her early biographers, Elizabeth Cleghorn

Gaskell, that Shirley is as Emily "would have been, had she been placed in health and prosperity."*

Of Charlotte's four novels, fully two boast rabies subplots. In *Shirley,* as we have seen, the disease serves as pretext for a hard-edged woman to marry; in *The Professor,* it serves as shorthand for a father's manly duty. The narrator (and professor of the title), William Crimsworth, describes his son by telling the story of when the boy's mastiff, Yorke, was bitten by a rabid street dog. As soon as the elder Crimsworth discovers this fact, he immediately shoots Yorke dead, not knowing that the son is looking on in horror. The boy proceeds to make a spectacle of his grief for weeks, even prostrating himself out on the dog's burial mound. Like *Old Yeller* and *To Kill a Mockingbird* in our own time, *The Professor* uses rabies above all as a means to establish a man's courage, to delineate his duty and his dominion.

In many ways, though, it is Charlotte's first and most famous novel, *Jane Eyre,* that captures best the zoonotic idea, the animalistic infection of which rabies serves as progenitor. When Jane becomes the employee, and later the fiancée, of Edward Rochester, her happiness and indeed her very life are threatened by a sinister presence in the attic, a madwoman whom we eventually learn to be Bertha Rochester, Edward's violently insane wife. Bertha's own brother, Richard Mason, is slashed and then savagely bitten by her in the night. Eventually, the creature appears to Jane in her room, soon before she herself is supposed to marry Edward. Jane recalls the encounter to him the following day:

---

* As it was, Emily would live to only the age of thirty, dying in December 1848 of what was probably tuberculosis. Her younger sister, Anne, died six months later of the same condition. Charlotte, the oldest, died at thirty-eight. But as Ann Dinsdale, librarian at the Brontë Parsonage Museum in Haworth, has remarked, "The surprise is not that the Brontës died so young but that they lived so long." A health report in 1850 found the life expectancy in Haworth to be just twenty-five. All six Brontë siblings lived through a bout of scarlet fever as children, a statistically unlikely occurrence; two-fifths of Haworth children perished before their sixth birthdays.

> "It was a discoloured face—it was a savage face. I wish I could forget the roll of the red eyes and the fearful blackened inflation of the lineaments! . . . [T]he lips were swelled and dark; the brow furrowed; the black eyebrows widely raised over the bloodshot eyes. Shall I tell you of what it reminded me?"
>
> "You may."
>
> "Of the foul German spectre—the Vampyre."

Although Bertha Rochester is not, the reader soon discovers, in any way a supernatural being, she nevertheless places *Jane Eyre* squarely in the nineteenth-century genre of monster tales—stories of humans who are not entirely human, who are tinged (indeed cursed) with some element of the animal or the bestial. In his book *Knowing Fear,* the horror scholar Jason Colavito charts the nineteenth-century rise in literature of what he calls "biological horror," featuring fully corporeal malefactors that "embody in their beings the struggle of humanity to re-imagine its relationship with the animal kingdom and the natural world." Thus the emergence of the *monster,* the non-man man, "a bizarre liminal creature poised somewhere on the continuum between man and beast."

It would be too much to credit rabies alone with the nineteenth-century boom in monster lore. But the (trumped-up) threat of hydrophobia did foment, and in turn exploit, the same visceral fear: that any proper middle- or upper-class man or woman, refined above any condition of existence that one could possibly consider base or animalistic, might suddenly and through no fault of his or her own be gripped by an insensate and subhuman savagery. This fear was well captured in an 1830 letter to the London *Times,* addressed from Boodle's, the tony gentlemen's club on St. James's Street, and genteelly bylined "A Constant Reader." "Who," asked this anonymous gentleman, during the height of that year's rabies outbreak,

> is there among us—either at the east or west end of the town—that can leave his home in the morning, and say that he may not return in a few hours, brought back in a state that would reduce him to the desperation and frenzy of a demon, and from which a horrible death can alone relieve him?

The man's unsubtle invocation of class ("at the east or west end"), nominally democratic, in fact serves the opposite function—pointing out that the most horrifying form of hydrophobia is the one that grips the man of means.

It is no accident that the most hysteria-inducing monsters of the era, factual or fictional, clambered their terrifying forms out of polite society. It was important that Dracula, like Polidori's Byronesque vampire, was an aristocrat in order for his murderous deeds to be truly chilling. The reader's shock at the actions of Robert Louis Stevenson's Mr. Hyde—"with ape-like fury," for example, "trampling his victim under foot and hailing down a storm of blows, under which the bones were audibly shattered"—was secondary to her shock that he was, in fact, the genteel Dr. Jekyll. Much the same was true of the real-life legend of Spring-Heeled Jack, a cloaked marauder, widely believed to be a nobleman, who attacked at least two young women in the London suburbs in 1837 and then persisted for decades as an English bogeyman, widely seen but never captured. After breathing fire into his victim's faces, he slashed at them with taloned hands and then escaped by leaping improbable distances.

That such monstrous transformations could be effected in even the most refined gentleman made clear the ease with which they might also befall oneself; as the aforementioned French doctor, G. E. Fredet, put it about rabies: "Leaving to the patient all the faculties of his intelligence intact, he sees himself die." Across the Atlantic, Edgar Allan Poe revolutionized the horror genre in part through his chilling use of first-person narration in describing just these sorts of descents into subhuman madness. In both "The Tell-Tale Heart" and "The Black

Cat," the most chilling dimension of Poe's murderous narrators is not the savagery of their crimes but the contrast between that savagery and their evident intellect, and the deterioration of that intellect in a way that strikes the literate reader as all too believable. "The Black Cat," in particular, is a conjuring trick, a narrative that begins in the mode of a sedate personal memoir ("From my infancy I was noted for the docility and humanity of my disposition") but soon sinks slowly into deepening depredations. First, the narrator mutilates his cherished cat: "the fury of a demon instantly possessed me"—he might as well have said "frenzy"—as he cuts an eye from the yowling animal's head. Then, as his madness deepens, and as his guilt over this misdeed compounds, he is driven (in "a rage more than demoniacal") to murder his wife with an ax and wall up her corpse in the cellar. Again, as Ovid demonstrated in his retelling of Actaeon's fate, the only thing more terrifying than the violence is a firsthand witnessing of the bestial metamorphosis that provoked it. These tales warn us that we are all vulnerable to the madness, capable of the savagery it inspires.

As it happens, Poe himself descended into madness in the hours preceding his death. Found unconscious on the streets of Baltimore on October 3, 1849, he was brought to Dr. John J. Moran of the Baltimore City and Marine Hospital. Though Poe was known for trouble with alcohol, the driver who picked him up swore to Moran the fallen man did not smell of drink. Soon, the author began engaging in (as Moran later described it in a letter) "vacant converse with spectral and imaginary objects on the walls"; Poe's "face was pale and his whole person drenched in perspiration." At times during his hospitalization, according to most accounts, Poe was calm and lucid. At others he was seized with "a violent delirium, resisting the efforts of two nurses to keep him in bed." Eventually, four days after admission, the great author expired.

In 1996, on a lark, a doctor at the University of Maryland Medical Center in Baltimore decided to put the case of Poe's death—stripped of his name and time period—before the center's weekly pathology rounds. Unaware that the patient in question was Poe, a cardiologist named R. Michael Benitez developed a theory that the most likely cause of death was rabies. As Benitez notes, the attending physician reported no signs of trauma and mentioned none of the extreme fever that one would associate with malaria or yellow fever. Strangest of all was Poe's cycle of relapsing, between periods of madness and periods of lucidity. The more obvious causes of death, especially the delirium tremens that might present in an alcoholic such as Poe, would have taken an inexorably progressive path, with his condition worsening steadily.

Rabies sufferers, by contrast, are prone to just these sorts of swings between delirium and lucidity. It's true that Moran made no mention of an animal bite, but as Benitez points out, of the thirty-three human rabies cases in the United States between 1977 and 1994, only nine presented with documented evidence of animal exposure. And the average number of days that neurological rabies sufferers survive is four: exactly the same number of days that elapsed between Poe's arrival at the hospital and his demise.

By the middle of the century, few physicians believed—as no less a figure than Benjamin Rush could in 1800—that humans sometimes acquired rabies spontaneously, without any contagion at all. But most doctors and veterinarians did still profess this belief about dogs. And to explain this supposed phenomenon, medical opinion had coalesced around a new, rather lurid theory: namely, that many if not most cases of dog rabies were caused by a lack of sexual satisfaction. The rise in pet keeping, in an era before the spay or neuter became common, focused nineteenth-century observers very uncomfortably on the unslaked sexual urges of their otherwise trustworthy companions. Dog owners confronted with the masculine fervor to mount during walks, or with the recurring frenzy of feminine heat, could be forgiven for later imagining that it was these unconsummated passions (and not the unseen nip from a stray in the streets) that had caused their pets to be seized by canine madness.

Henry William Dewhurst, already a figure of murky scientific standing (by the 1840s he would be denounced in medical journals as an incorrigible quack), propounded this theory in a thoroughly dubious 1830 address to the London Veterinary Medical Society. To support his argument that rabies could be spontaneous, he relayed the details of two cases—a terrier confined by an old lady, and a doctor with a sporting hound—in which dogs displayed symptoms of rabies but then recovered. As for his theory that lack of sex was to blame, he fell back on a general observation that when the passion for sex "is unable to be gratified, as was intended by the great Author of nature, pure madness breaks out." Here his lone example was the story of an elephant, kept in confinement, that had to be put down on account of its "unrestrained fury," though he neglected even to specify how sex figured into the tale.

Regardless, Dewhurst's idea gained great purchase in Victorian England, and the French, perhaps predictably, agreed. In her splendid survey of nineteenth-century pet keeping in Paris, *The Beast in the Boudoir,* the historian Kathleen Kete cites one generally respected text from 1857 that posits sexual frustration as the *sole* cause of rabies in the dog. (Its authors, the doctors F. J. Bachelet and C. Froussart, lament that dogs do not have the same recourse to self-satisfaction that humans have.) That this theory had found purchase in Italy, too, is evidenced by an 1845 proposal, penned by a certain Monsignor Storti under the title "Project for the Prevention of Hydrophobia in Man," detailing the creation of what can only be described as mandatory canine bordellos. Under this plan, each male dog would be brought to a central location for his urges to be gratified. Immediately afterward, he would be neutered and then sold. And then—presumably in order to keep these dogs from generating rabies eventually—all male dogs would be destroyed two years after their sale.*

---

* "We have little desire to disturb the dream of a benevolent man," commented the *Gazzetta Medica di Milano* on the publication of his proposal. "We cannot, however, help stating that, while reading over his plan, a slight difficulty occurred to us.

Unlike Bachelet and Froussart, most believers in spontaneous generation did not see sexual frustration as the only cause. Dehydration was another commonly named culprit: because dogs do not sweat, and instead regulate their temperature through panting, medical men of the era thought that intense thirst on hot days could prompt an animal's blood to fester into a poisonous state. Similarly, it was thought that rabies could results from dogs' exposure to their own excrement, through contact with it—or, worse, consumption of it.

Regardless of particular theories, this belief in spontaneous generation of rabies had dire effects in the public-health battle over hydrophobia. In 1830, when the British Parliament was crafting a bill to contain the disease, members heard testimony from two noted veterinarians, one of whom believed in spontaneous generation, the other of whom did not. Naturally, these differing beliefs led to two very different prescriptions: the former held that confining dogs would alleviate the epidemic, whereas the latter testified that it would have the opposite effect, generating new cases as confined dogs became "impregnated with an animal poison from the lungs, faeces, urine, and skin." Similar battles were fought over the mandatory use of muzzles, another effective containment strategy. No consensus could be reached on these legislative issues so long as the science of rabies remained unsettled. And as late as 1874, with Pasteur's cure only a decade away and the germ theory of disease already percolating through the medical community, a tally of rabies experts surveyed by M. J. Bourrel, the former chief veterinarian of the French army and a staunch contagionist, found that believers in spontaneous generation far outnumbered his own side.

Meanwhile, the burgeoning emphasis on the sexual nature of

Suppose the establishment [is] in operation and flourishing. All dogs have been killed by their masters, all canine importation has been prohibited, and, lastly, all the new-born in the *seraglio* have been pitilessly castrated. So far [so] well! But what dogs remain to frequent such establishments? Where are new recruits to be found?"

rabies in the dog coincided with a growing emphasis on the more lurid dimensions of the fatal illness in humans. It was not until the nineteenth century that priapism, satyriasis, and nymphomania in women began to appear in all standard lists of hydrophobia symptoms. Often these mentions were accompanied by a thirdhand anecdote about some man who, like Galen's porter, was given over to extreme sexual release in his final hours. The most common such tale, about a man who ejaculated thirty times in one day, originated with the eighteenth-century doctor Albrecht von Haller. In the retelling, this became recast as a bravura act of performance—for example, "that rabid man related by Haller, who accomplished the sexual act 30 times in 24 hours"—as if the dying man had wooed a new and willing partner for each. In their book on rabies, Bachelet and Froussart emphasize the weakness of women, too, to these depredations, claiming that a female equivalent of satyriasis (*fureur utérine*) is apparent in autopsies of rabies victims. They go on to describe nymphomania itself as a related and similarly fatal condition whose path is nearly identical to that of hydrophobia:

> As the symptoms intensify, a *frothy foam* dribbles from the lips, the victim's breath becomes fetid and her thirst, burning. Often these symptoms are accompanied by an intense fear of water . . . , *gnashing teeth,* the *desire to bite,* and death is not slow in putting an end to these horrible afflictions.

It seemed that rabies hysteria was reaching its historical apex—ironically, at the very same time as medicine stood on the cusp of a rabies cure.

Out on the vast American plains, rabies stalked the nineteenth-century frontiersmen in a form unimaginable in Europe, though no less diabolical. Indeed, Canadian trappers came to bestow on the offending creature the dramatic nickname *l'enfant du diable,* or "the child of the devil," a fitting name given its foul odor and sinister black coat. Americans called it something more prosaic: the "'phoby cat," that first word being short for "hydrophobia." The malefactor in question was the skunk.

It was widely believed among Plains travelers that skunk bites universally led to rabies infections. No less an American than Theodore Roosevelt wrote that "there is no wild beast in the West, no matter what its size and ferocity, so dreaded by old plainsmen as this seemingly harmless little beast." In the 1870s, when the army colonel and memoirist Richard Irving Dodge was commanding one frontier fort, all sixteen cases of reported skunk bite led to fatality; at another fort, the surgeon put the death rate at ten cases out of eleven. The lesson of those startling numbers, obviously, is not that all skunks were carrying the disease but rather that skunks do not attack (or even approach) humans except when in the demented throes of rabies.

Roosevelt, in one of his memoirs, recalled a hunting trip when a skunk encounter went comically awry. A hungry 'phoby cat burrowed under the wall of their log hut in search of food. Despite the close quarters, one of the hunting party—Sandy, a "huge, happy-go-lucky Scotchman"—shot at the creature with his revolver, waking his huntmates with a terrible jolt and causing general consternation among them; the shot didn't injure any hunters, luckily, though it didn't hurt the skunk, either. Half an hour later, the skunk returned, and, recalled the future president, "the sequel proved that neither the skunk nor Sandy had learned any wisdom by the encounter": Sandy shot again, at which his sleeping companions jumped up and fled the hut in their confusion. (This time, however, Sandy hit the skunk. "A did na ken 't wad cause such a tragadee," TR reports the Scotchman having said, morosely.)

L'enfant du diable was not the only New World animal to harbor rabies. Less common, though arguably more fearsome, was the rabid wolf. Perhaps the most spectacular attack on the Great Plains occurred in 1833, along the Green River in what is now western Wyoming. That summer, multiple teams of fur traders, led by the Rocky Mountain Fur Company, had an appointment to meet for one of their semi-regular "rendezvous." In mid-July, the encampments of these traders were terrorized by what one nineteenth-century chronicler called "one of those incidents of wilderness life which make the blood creep with horror." A raving wolf tore along the river, through the camps of sleeping traders, biting them and their cattle. In one camp, he was said to have bitten twelve men. In another, an eyewitness reported that three men were bitten in their tents, all in the face. It's unknown how many men died from their wounds; some of the accounts, like that of the Indian who "shortly afterwards" began to "roll frantically on the earth, gnashing his teeth and foaming at the mouth," or the trader who threw himself from his horse and began "barking like a wolf," sound suspiciously exaggerated. But most accounts confirm that men and livestock came down with the disease.

Another rabies-addled wolf rampaged through Kansas's Fort Larned in 1868. In one of his memoirs, Colonel Dodge passes along a full account of the incident, as put down in the fort's own records: "On the 5th August, at 10 p.m.," the account ran,

> a rabid wolf, of the large grey species, came into the post and charged round most furiously. He entered the hospital and attacked Corporal, who was lying sick in bed, biting him severely in the left hand and right arm. The left little finger was nearly taken off. The wolf next dashed into a party of ladies and gentlemen sitting in Colonel's porch, and bit Lieut. severely in both legs.
>
> Leaving there, he soon after attacked and bit Private in two places. This all occurred in an incredibly short space of time; and, although those abovementioned were the only parties bitten, the animal left the marks of his presence in every quarter of the garrison. He moved with great rapidity, snapping at everything within his reach, tearing tents, window-curtains, bed-clothing, &c, in every direction. The sentinel at the guard-house

> fired over the animal's back, while he ran between the man's legs. Finally he charged upon a sentinel at the haystack and was killed by a well directed and most fortunate shot.

Frederick Benteen, who would later become famous (and, to some, infamous) for his disobedience of Custer's orders at Little Big Horn, was stationed at Larned at the time of the attack, and he recalls that all the bitten soldiers, save one, died of hydrophobia. That one, a soldier named Thompson, "was saved on account of wolf biting through pants, drawers and socks, thus getting rid of all the virus on clothes," Benteen wrote to a correspondent in 1896. "It scared Thompson 'pissless,' as we say in the cavalry, and well it might!"

Rabies cures on the frontier were a mishmash of medicine and folk remedy. The benefits of cauterization and bleeding, those brute-force but nevertheless somewhat effective tools of the ancients, were generally known, though quack treatments like nitrate of silver were also sometimes applied. One novel cure, recommended by one Western commentator, was the use of progressively greater quantities of the deadly poison strychnine.

But there was also a particular fascination with Native American cures. The anthropologist George Bird Grinnell, recounting his time with the Blackfeet, describes a cure for rabies whereby the sufferer was essentially sweated out of his illness. His relatives bound his hands and feet, rolled him in a buffalo hide, and built a fire not just around him but on top of him as well. The natives explained that "so much water [came] out of his body that none was left in it, and with the water the disease went out, too." Colonel Dodge put forward the truly odd claim that skunk bites, so fatal to the white settlers, did not affect Native Americans one bit, whereas for wolf bites the situation was reversed: "In every instance, death by hydrophobia is to the Indian the sure result of even the slightest scratch from the teeth of the rabid animal." (Dodge adds: "They make no attempt at treatment, but philosophically commence preparations for the death sure to come in a few days.") Of one particularly tantalizing Native American cure for rabies, the existence is recorded but the specifics lost to history. In 1827, the War Department went so far as to solicit federal agents in Indian country to ascertain what cure the Native tribes had for hydrophobia. Thomas McKenney, of the Office of Indian Affairs, received a detailed reply from the field, describing a medicinal plant that the natives claimed would cure rabies. It's unclear whether the government tested its efficacy, but the Department of War was clearly serious in hoping that white farmers would learn to cultivate the crop as a cure: when McKenney passed along the correspondence to the *American Farmer* magazine, he also sent the magazine a packet of seed, "with the view to have it distributed, in your discretion, for the preservation and multiplication of the plant."

Interest in these Native cures seemed to flow from the presupposition that these so-called Indians themselves lived in a sort of animal state: settlers felt as if the wolfish disease should be best understood by those who were closer in nature to the wolf. This identification of Native Americans with wolves dated from the earliest Pilgrims, who had arrived with their own superstitious beliefs in the wolf as an evil, almost supernatural predator. As Jon T. Coleman, a historian at Notre Dame, points out in his book on wolves in early American culture, *Vicious:* "From the colonists' perspective, Indians sang, talked, prayed, fought, and traveled like wolves." In 1642, the Massachusetts governor, John Winthrop, described the newly settled land as overrun with "wild beasts and beast-like men." Later, an eighteenth-century clergyman in Northampton, Massachusetts, in decrying the guerrilla warfare of the natives, noted that they "act like wolves and are to be dealt withal as wolves." It should be noted that many of the natives associated strongly with the wolf themselves. One tribe, the Skidi Pawnee, dressed in cloaks of wolf skin and were known as the Wolf People by other tribes in the region; individuals in many tribes took names that claimed some kinship with the wolf.

Regardless, as wolf and native both were beaten back over centuries of brutal eradication, the frontier attitude toward both seemed to soften—from outright hatred and fear to a sort of colonial condescension. More typical, over time, was the attitude toward wolves struck by Francis Parkman in his 1849 book about the Oregon Trail: "There was not the slightest danger from them, for they are the greatest cowards on the prairie."

In the waning days of the nineteenth century, the key weapons that would be required to slay rabies were actually being forged in the study of a different disease—the study, in fact, of the one other illness generally known at that time to pass from animals to humans. Anthrax was an ideal target for the early adherents of germ theory, because it was a spore-forming bacterium, rather than a virus, and as such it is abnormally large for a pathogen. Moreover, unlike many of its microbial brethren, it is found in copious numbers in the blood of late-stage patients.

Isolation of the anthrax bacillus was accomplished by the great German physician Robert Koch, who was not even thirty, and just a country doctor, when he took a position as a local medical official in the town of Wollstein and began carrying out his pioneering research. Through the pinching of pfennigs (in particular, by forgoing the purchase of a carriage, which he would have needed to perform house calls), Koch was soon able to acquire a microscope; he chose one made by Edmund Hartnack of Potsdam, in Germany's east, arguably the finest microscope maker of the day. Koch began his studies of anthrax in 1873, and by Christmas 1875 he had not only definitely discovered the microbe responsible but also tracked its entire life cycle in a rabbit. More important still, he had learned to culture the microbe artificially, by using the aqueous humor of the rabbit's eye as a sterile medium. The resultant paper—"The Etiology of Anthrax, Based on the Life History of *Bacillus anthracis*," published in 1876, when Koch was only thirty-two—helped both to establish the field of microbiology and to establish its author as one of that field's foremost practitioners.

Microbiology's other titan was twenty years older and worked three hundred miles to the west, in Paris. While Robert Koch's intellectual pleasure lay in pure discovery, Louis Pasteur was obsessed with practical applications. After creating a procedure for removing harmful microbes from milk and beer (which we still know today as pasteurization), he had turned his energies to inoculation. By 1876, he had already developed a vaccine for a disease in poultry, and after reading Koch's paper, he turned his attention to anthrax. In his successful creation of an anthrax vaccine, he honed the method—attenuation, or the weakening of live pathogens—that he would soon apply to the even more insoluble-seeming problem of rabies. Four thousand years had elapsed since the Laws of Eshnunna warded against the rabid dog; but over the course of just five years, from 1880 to 1885, both our ignorance and our terror of rabies would be entirely upended by the work carried out in one laboratory on the rue d'Ulm.

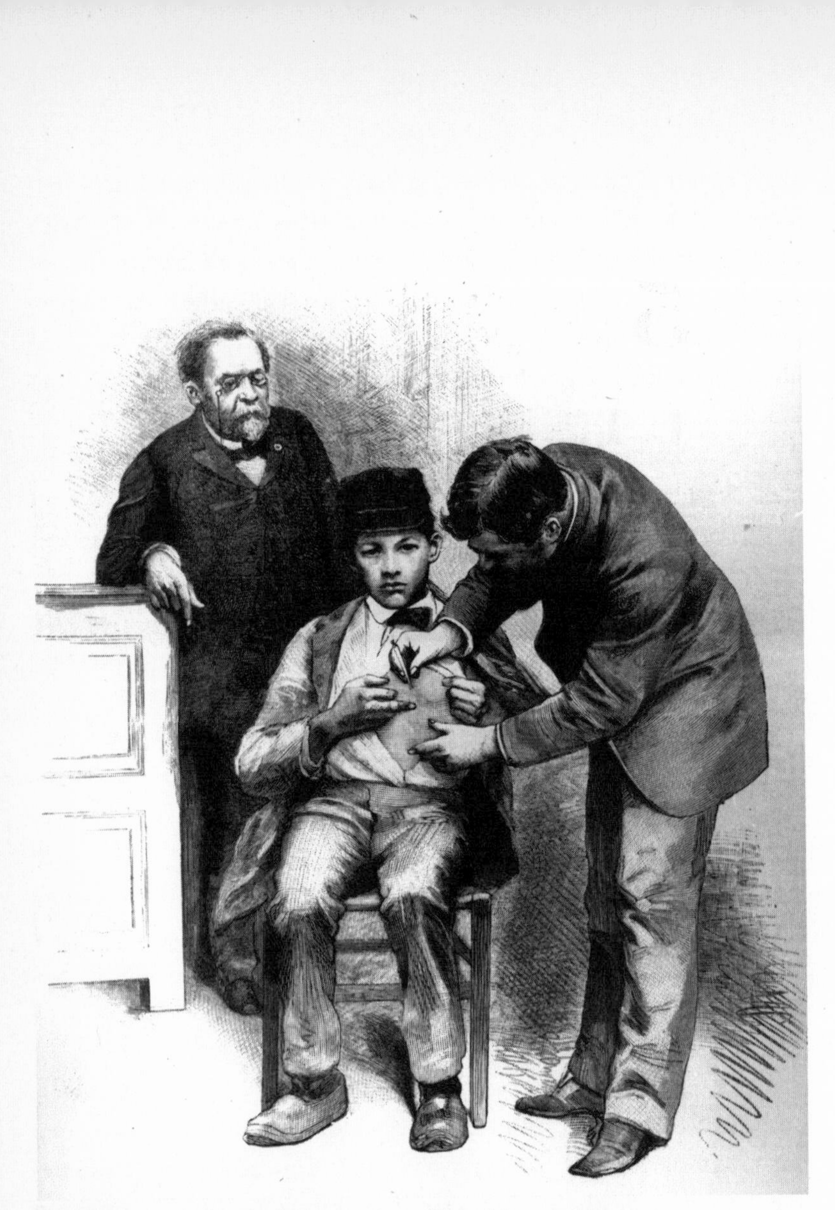

Louis Pasteur oversees administration of the rabies vaccine.
Cover of *L'Illustration* magazine, 1885.

5

KING LOUIS

He was born in 1822 to a father who was a provincial tanner of animal hides—a bitterly nostalgic veteran of the Napoleonic Wars who would pin his Legion of Honor ribbon to his spotless frock coat for his regular solitary Sunday stroll through town and field—and to an imaginative, enthusiastic mother from a large family of warmhearted gardeners. Their snug little home above the tannery in Arbois was thick with the fetid smell of wool grease. Still, Louis Pasteur's rustic boyhood was a happy one. He enjoyed fishing, sledding, and the company of his three sisters. An eager but undistinguished student, he was noticed primarily for his artistic abilities. While some of the locally distinguished acquaintances who sat for portraits with the young Pasteur supposed he might someday have a modestly bright future in painting, his father's ambition was that Louis would eventually achieve the respectable station of secondary-school instructor.

Pasteur was not quite nine years old when his quiet village childhood was punctuated by a disturbing drama. After hearing reports of a rabid wolf marauding through the region of Arbois, furiously biting man and beast, Pasteur and his friends witnessed one victim being brought to the blacksmith's shop for treatment. The sight of a red-hot iron cauterizing the man's still-frothy wounds made a lasting impression on the young Pasteur. So did the hydrophobic deaths soon afterward of eight of his fellow Arboisians, who had suffered the wolf's bites on their hands and heads.

As a university student in Paris, Pasteur was exposed to the great scientific minds of his day, and his own gift for original research finally surfaced. Following completion of his master's degree at the École Normale Supérieure, Pasteur chose not to return to the provinces as a schoolteacher and instead took a position in the laboratory of the famous chemist Jérôme Balard. He defended theses in physics as well as chemistry and after one year's time made his first report to the French Académie des Sciences on the relationship between various crystalline forms of particular chemical substances and the rotational polarization of light—a paper that elegantly unified much of the contemporary research into molecular physics and chemistry.

Pasteur was appointed a professor of chemistry at the University of Strasbourg, where he underwent two important transformations. First, almost immediately upon meeting Marie Laurent, the gentle, patient daughter of the university provost, Louis Pasteur began a campaign to win her hand that would soon seal his fate as a dutiful family man. Second, while he continued to make discoveries in the laboratory concerning the physical and chemical character of crystalline substances, Pasteur increasingly concerned himself with scientific problems that had direct practical applications, such as the process for industrial production of racemic acid crystals and, later—as the dean of the faculty of science at Lille—the fermentation of beetroot alcohol. Pasteur saw himself as performing science for the people: the French people in particular. Before long he would likewise be known as the people's scientist.

In 1857, Pasteur returned to the École Normale in Paris. Here his ongoing discoveries in the fermentation and spoilage of wine, which he established to be microbiological processes, led to studies that exploded the stubborn myth of spontaneous generation and notably led to the development of new preservation methods for perishables. Pasteurization was born and would change forever the handling of food and drink. Pasteur immediately and doggedly began to explore the relationship between the putrefaction of foodstuffs and the necrosis of diseased tissues. As his research interests evolved from physics and chemistry to microbiology and medicine, the general populace became increasingly interested in his work. Indeed, the emperor and empress themselves began to pay keen attention, and Pasteur, for his part, quickly learned how to transform public interest in his research into material support for the glassware, incubators, laboratory bench space, animals, and capable assistants his ongoing endeavors would require.

Many scientists would train and toil by Pasteur's side, but the flinty young physician Emile Roux contributed more than any other to Pasteur's researches into animal and human disease processes. Roux's medical training had been temporarily disrupted when he, in a fit of anger at the director of the Val-de-Grâce Army Medical School for slighting the serious scientific effort he was expending on his student dissertation on rabies, insulted his superior and was consequently imprisoned and then expelled. As a medical graduate, Roux monastically devoted himself to the systematic study of microbes and immunity in the Pasteur laboratory at the École Normale (and later at the Institut Pasteur). In this role, he was often a thorn in the side of his master, pitting his own methodologies against Pasteur's, always urging the elder scientist toward greater extremes of scientific rigor. "This Roux is really a pain," Pasteur complained. "If you listened to him, he would stop you in everything you are trying to accomplish." Still, the collaboration between the two men, which lasted from 1878, when Pasteur began to concentrate on contagious diseases, until Pasteur's death in 1895, was an extraordinarily productive one.

Throughout his career, Pasteur was known for his diligence and tenacity: he would approach every research question with an exhaustive, meticulous zeal. Since he often took on problems of particular controversy in his own lifetime, his rigor was never wasted, as he was constantly under attack. The French scientists of the nineteenth century were not content to air their disagreements in sternly worded missives placed in relevant academic journals, as is standard today. Rather, they confronted one another face-to-face before their esteemed colleagues at the Académie des Sciences, the Académie Nationale de Médecine, or even the venerated Académie Française. Pasteur's fastidious methodology was matched by his aggressive rhetorical manner, a combination that frequently allowed him to make a great show of toppling his rivals' scientific theories in public—indeed, to terrific applause. These performances were a particular source of satisfaction for Pasteur.

Pasteur professed a belief in research and experiment as a means to end human misery. It was a goal both lofty and earnest. He advised his younger scientific colleagues, "Live in the serene peace of laboratories and libraries. Say to yourselves . . . , What have I done for my country? Until the time comes when you may have the immense happiness of thinking that you have contributed in some way to the progress and to the good of humanity." Pasteur's love for children, in particular, and passion for preserving them against the morbid threat of infectious disease were to become famous. "When I see a child," said Pasteur, "he inspires me with two feelings: tenderness for what he is now, respect for what he may become hereafter." Much of Pasteur's medical research focused on diseases that were particularly associated with childhood illness. Pasteur himself had lost three children to disease: two young daughters to typhoid and another to cancer. Following the death of the third daughter, Cecile, he wrote to a colleague, "I am now wholly wrapped up in my studies, which alone take my thoughts from my deep sorrow."

Pasteur felt his calling as a scientist was ultimately to spare life and alleviate suffering, and as the secrets of microbiology revealed themselves to him over the course of his career, his conscience guided him toward new humane applications. Early in his career, he painstakingly tested and vigorously defended techniques to control microbial contamination—not just of food and drink but also of surgical wounds—and in doing so saved countless lives around the world. But as his restless mind turned toward other diseases, contagions first of France's livestock and then of its countrymen, Pasteur began to think of a more fundamental means of preventing the morbidity and mortality caused by infection. Vaccines took hold of his imagination.

Vaccination is the induction of immunity to a disease in an otherwise vulnerable individual, accomplished through intentional exposure to some less virulent form of the disease. The practice began with variolation against smallpox infection, originating in Asia more than a millennium before Pasteur. The procedure consisted of taking a small amount of the pus from an active smallpox lesion and introducing it into a small surgical incision or directly into the nose of a patient with no history of the disease. The resulting infection was milder and less disfiguring than natural smallpox, leading to a case fatality rate of only 1 to 2 percent, as opposed to 30 percent with a natural infection. Variolation was popularized in western Europe during the early eighteenth century by England's Lady Mary Wortley Montagu. After witnessing its successful practice during her husband's term as British ambassador to the Ottoman Empire, Lady Montagu insisted that her three-year-old daughter be variolated a few years later when a smallpox outbreak threatened England. Much interest was generated in the British court, and within a year the Prince of Wales's daughters Amelia and Caroline had been variolated as well. The practice immediately became widespread throughout Britain but had yet to overcome several decades of resistance in France. Only after the unexpected death of Louis XV from smallpox in 1774 did variolation become common among the French.

Since variolation was neither affordable nor accessible to the lower classes, it was never taken up generally as a preventative. Instead, large-scale vaccination efforts were set up only after an epidemic was in progress, limiting the ability of variolation to make a broad impact against smallpox. The physicians who carried out these procedures had no genuine scientific knowledge of why they were effective; it would be more than a century before Pasteur would popularize the germ theory and establish microbiology and immunology as fundamental medical sciences. Many physicians of the eighteenth century believed, for example, that variolation was most survivable when combined with fasting, bleeding, and mercurial purges.

One British country physician, still stricken by the memory of the noxious variolation protocol he himself underwent as a child, was motivated to find a way to diminish danger and discomfort to the patient without compromising protection against the dreaded smallpox. And so Edward Jenner set out to test the folk belief that those who handle cattle from a young age, and thus have the opportunity to be exposed to cowpox, or vaccinia, before they encounter smallpox, are protected from smallpox infection. Once this hypothesis was confirmed, he demonstrated that vaccinia could be intentionally inoculated into a naive human, conferring similar protection. His simple experiments on his neighbors and family members sufficed to convince the world that rather than being variolated with potentially deadly active smallpox, one could be inoculated with a much less virulent disease associated with altogether different species and thereby be protected from the graver infection. This humane innovation was quickly taken up around the globe; more than 100,000 were vaccinated before ten years had passed, and Jenner became an international celebrity. Immediately upon the creation of vaccine came the birth of the antivaccine movement, scientists and laypeople who claimed (much as in our present day) that vaccine was "poison." But its use became more and more widespread, even compulsory in many places, as the decades wore on and vaccine production became standardized and improved.

Altogether, it would take less than two centuries for Jenner's vaccine to eradicate the scourge of smallpox from the earth.

Louis Pasteur favored preventative strategies against infection, and he was a great admirer of Jenner and the principle of vaccination. By the time Pasteur began his own work on communicable diseases, Jenner's legacy was firmly established, if still not well understood. The Académie Nationale de Médecine recommended general vaccination but was still struggling to differentiate the agent of the vaccine from that of smallpox itself. Pasteur's interest extended well beyond smallpox. He was determined to figure out the general method for immunizing patients against all of the different pathogenic microbes being cultivated in his laboratory.

Chicken cholera was the first disease to yield its secrets to the Pasteur research team. This bacterial disease of fowl was rampant in France during the 1870s, bringing misfortune to poultry farmers across the countryside. According to one of Pasteur's assistants, Émile Duclaux, the breakthrough was made after the culturing of microbes was interrupted for the summer holiday. When the new academic year commenced, it was noted that the bacteria that had been set aside no longer transmitted the disease.* The formerly deadly germs produced no grave effect on experimentally infected healthy chickens. Intrigued, Pasteur took these same chickens and submitted them to a second experiment, alongside chickens that had never received inoculations. He infected both groups of animals with very fresh chicken cholera isolates, of determinately high virulence, and monitored them closely

---

* Duclaux's account was supported by the biography written by Pasteur's son-in-law, René Vallery-Radot, but some modern scholars dispute it. In 1985, based on a thorough study of Pasteur's notebooks, the French historian Antonio Cadeddu asserted an alternate history: Pasteur's collaborator Roux determined the method for attenuation of chicken cholera through prolonged, deliberate laboratory experiment—without the knowledge of Pasteur.

for ill effect. Shortly, Pasteur was able to observe that the birds exposed originally to the aged bacteria resisted infection with the virulent strain, too, while the naive chickens succumbed.

The significance of this finding was not lost on Pasteur. Here was induced immunity from a mortal disease—not happened upon fortuitously in the cowshed like Jenner's, but experimentally produced in the laboratory! If chicken lives could be spared through inoculation of laboratory-attenuated microbes, it did not require much imagination on Pasteur's part for him to suppose his method may have potential for saving human lives as well. As a nod to Jenner, Pasteur referred to his method of chicken-cholera immunization as a "vaccine."

Pasteur's new vaccine soon attracted naysayers on several fronts: those who fought against all science based upon the germ theory; the anti-vaccinists (who had already honed their rhetoric against the Jennerian vaccine); and those scientific rivals who would have invented the chicken-cholera vaccine themselves if their own methodology had been more sound. Pasteur was in the midst of preparing his findings for the Académie Nationale de Médecine when his arguments with his rivals in that body became so heated that he received an invitation to duel from the aging surgeon Jules Guérin. (The sixty-year-old, hemiplegic Pasteur was delicately extricated from the challenge by friends in the Académie.)

To test the broader utility of his method, Pasteur turned his attention to a second veterinary disease, one with greater economic importance for French and European agriculture: anthrax. While capable, in rare instances, of dealing a farmer or veterinarian a grisly death, anthrax was most feared across rural Europe for its ability to depopulate a prosperous farm, leaving behind an indefinitely contaminated field. Spurred on by Robert Koch's pioneering paper, Pasteur set out to attenuate the isolated anthrax bacillus in a similar manner as he had done with chicken cholera. He was soon successful: after achieving partial success with heat deactivation, the Pasteur team ultimately found that temperamental anthrax was best attenuated chemically, with carbolic acid treatment.* In the end, the pathogen proved no less amenable to laboratory domestication than chicken cholera had.

The furor among France's scientists and medical men created by Pasteur's announcement of the anthrax vaccine was so intense, so fevered, as to demand some public proof of his claims. The influential veterinarian Hippolyte Rossignol accused Pasteur of "microbiolatry" in an editorial in his *Veterinary Press.* He invited Pasteur to perform a public demonstration on Rossignol's own Pouilly le Fort farm in the pastoral Brie region east of Paris. Pasteur accepted the challenge, eager for a means of advancing his doctrine of vaccination. He devised a simple experimental protocol: twenty-five sheep would be vaccinated against anthrax, fifty including these would be infected, an additional ten would serve as untreated controls. All sixty would be monitored for subsequent ill health. The demonstration, carried out during May 1881, was witnessed in its various stages by a large rabble of farmers, physicians, pharmacists, newspapermen, and, especially, veterinarians—many of whom remained as skeptical of Pasteur's vaccine as they were of the germ theory that gave birth to it. Far away from the pasture where the vaccine trial took place, some of Europe's most ardent germ theory supporters, Robert Koch and his assistants, suspicious that Pasteur's strong public assertions regarding microbial attenuation rested on as-yet-unstable science, voiced their stern disbelief as well.†

Great excitement was focused on the final stage of the trial, when the vaccinated and unvaccinated groups would both be injected with virulent anthrax. At the last-minute insistence of one of the more passionately skeptical veterinary observers, a triple dose of live anthrax was administered to each of the experimental animals. Pasteur

---

* The superior technique of carbolic acid attenuation was devised by the veterinary researcher Henri Toussaint and perfected by Pasteur's assistants, Roux and Charles Chamberland, after Pasteur had already announced the creation of attenuated anthrax using temperature manipulation.

† The Koch group, which relied on different culture methods than did the Pasteur laboratory, doubted the particulars of Pasteur's thermal method of attenuation.

himself vigorously shook the vial of anthrax prior to each inoculation, in order to prevent uneven distribution of the virulent principle. Other veterinary spectators demanded that the injections proceed with careful alternation between vaccinated and unvaccinated subjects. Pasteur assented indifferently to the various dictates of the veterinary crowd, never backing down from his assertion that "[t]he twenty-five unvaccinated sheep will perish; the twenty-five vaccinated ones will survive."

Pasteur projected complete confidence but was privately anguished as he waited to learn the fate of the herd. As the hours ticked by and the only news from Rossignol's farm was of a sick ewe from the vaccinated group, Pasteur's resolve began to waver. But two days after the inoculation, all twenty-five of the unvaccinated sheep were dead, while all of the vaccinated sheep had survived. "As M. Pasteur foretold at two o'clock 23 sheep were dead," the *Times* of London observed. "Two more died an hour later. The sheep which had been vaccinated frolicked and gave signs of perfect health. Farmers now know that a perfect prevention exists against anthrax."

Pasteur was roundly congratulated, especially by France's veterinarians, who had become his newest allies—allies who would prove extremely useful as his research progressed into the most fearsome disease known to that profession.

From anthrax, Pasteur turned his attention next to another veterinary disease, but one with widely understood consequences for people. Rabies, and its associated illness in humans, hydrophobia, did not claim so many French lives as others did. That said, it had a prominent place in the French imagination. For each one of the few hundred deaths from rabies registered each year in France, there were several bitten Frenchmen—or, more frequently, French children—who, along with their loved ones, spent many months in the agony of uncertainty: Would the wound lead to a grisly death from hydrophobia? In Pasteur's youth, when his own village had been terrorized by the rabid wolf, the danger was viscerally understood. But even as the scientist aged, the debate about whether rabies was a contagion or a spontaneous occurrence raged on among France's prominent biologists, physicians, and veterinarians.

Pasteur's collaborator Roux believed that Pasteur selected rabies as a subject for research as a calculated bit of stagecraft, so that his ideas about vaccination would attract maximum public interest. "This malady is one of those that cause the smallest number of victims among humans," Roux later wrote. "If Pasteur chose it as an object of study, it was above all because the rabies virus has always been regarded as the most subtle and the most mysterious of all, and also because to everyone's mind rabies is the most frightening and dreaded malady. . . . He thought that to solve the problem of rabies would be a blessing for humanity and a brilliant triumph for his doctrines."

The Pasteur laboratory received its first mad dogs from M. J. Bourrel, the former army veterinarian whose 1874 survey had found the anti-contagionists ascendant. Bourrel had been studying rabies for some years without penetrating very deeply into its mysteries. He had, however, localized its contagious principle to the rabid animal's saliva; given this fact, he recommended the precautionary measure of filing down the teeth of all dogs at large, so that should they become infected, they might not be able to penetrate skin and transmit the deadly agent. Bourrel could provide no better preventative than this, as his search for a rabies cure had led nowhere. He wrote in 1874 that his efforts in the laboratory had shown only that rabies is "impenetrable to science until now." In the summer of 1880, while assisting him in the laboratory, Bourrel's own nephew suffered the bite of a rabid dog and died following several days of torturous agony.

In December of that year, Bourrel provided the Pasteurians with two terrifying specimens of canine rabies for study. The first suffered from dumb, or paralytic, rabies. Its mute affliction was wretched to witness: a paralyzed, slack jaw, failing to support a limp, foam-covered tongue, and, above this, eyes full of "wistful anguish." The second dog, a victim of the more common furious form of the disease, terrorized the laboratory, menacing the scientists with its enraged, bloodshot gaze, its unpredictable lunges and fits, and its unforgettably mournful, hallucinatory howls.

During the same month, a doctor named Odilon Lannelongue contacted Pasteur about a five-year-old patient, bitten on the face one month prior to hospitalization, now racked by all the classic symptoms of rabies: restlessness, convulsions, aggression, hydrophobia. The child suffered mightily for fewer than twenty-four hours in the hospital and then died, his mouth full of the viscous mucus he had been unable to swallow. Within four hours after the child's death, Pasteur collected a sample of the mucus. Upon his return to the laboratory, he inoculated some of the diluted mucus into a group of rabbits—a procedure, published more than a decade earlier by the veterinarian Pierre Victor Galtier, proven to determine whether rabies was present in the saliva of suspect dogs. But the rabbits inoculated with the child's mucus surprised Pasteur, and contradicted precedent, by dying too quickly: they died in only thirty-six hours, when it should have taken weeks. Rabbits inoculated with saliva from those dead rabbits died nearly as rapidly. Moreover, the rabbits died of apparent respiratory failure, not neurological disease as would be typical of rabies. Dr. Lannelongue and his colleague Dr. Maurice Raynaud, after repeating the experiment themselves, eagerly announced proof that the child had died of rabies. If they were correct, this also would represent the first documented case of human-to-animal transmission of the disease.

Pasteur did not commit himself. He cultured a figure-eight-shaped microbe from the blood of the dead rabbits in veal broth and tested its virulence in more rabbits and also in dogs. Again, it swiftly dispatched its recipients. With further investigation, Pasteur and his assistants found that they could isolate and culture this organism from patients who were hospitalized with illnesses completely different from rabies—even from healthy adults. Pasteur named the microbe pneumococcus and declared that he was "absolutely ignorant of any connection that there may be between this new disease and hydrophobia."

Critics seized on this as evidence of the slipperiness of germ theory. Pasteur claims to be working on one disease, they sneered, but instead is working on another. To this notion Pasteur responded indignantly, "This is indeed a new disease produced by a new microbe; neither the microbe nor the disease has been described before. This tenacity in research, Monsieur, is the honor of our work, and it was because we, my collaborators and myself, pursued these experimental combinations that we were able to demonstrate that the new disease existed in the buccal mucus of children who had died of the same disease as well as in the saliva of perfectly healthy persons. It was then, and only then, that I had the right to assert that the new disease had no relation with rabies."

If rabies was not pneumococcus, then what was it, exactly? Despite a thorough investigation using all the tools of the Pasteur laboratory, no combination of methods and media available to Pasteur and his assistants would yield a microbial cause for rabies. Even as Pasteur's team discovered that the infectious principle for rabies resided in the central nervous system as well as in the salivary glands, they failed to culture a pathogen from either location. Thanks largely to the work of Pasteur himself, it was by this time a basic tenet of medical science that infectious diseases are caused by specific demonstrable microorganisms. Robert Koch's famous "postulates," first articulated in 1880, had made clear the relationship between microorganisms and disease, defining a disease-causing microbe as one that appears exclusively in diseased individuals; that can be isolated and cultured from a diseased host; that will cause disease when next introduced into a susceptible host; and that can be subsequently recovered from the experimental host and shown to be identical in culture to the microbe originally isolated. For rabies, not a single one of these conditions had been met. Koch's precepts have often been summed up with the phrase "one disease, one microbe," and Pasteur concurred with this view, but his vision saw a third term in this equation: one vaccine. He believed that every disease-causing microbe, once isolated, could be attenuated so as to safely confer immunity on a potential host. But it was hard to see how this equation could hold true unless a pathogen could be isolated, identified, trapped under glass, and then tamed.

Pasteur referred to the unseen—and apparently unseeable—agent of rabies as a virus. As his biographer Patrice Debré observed a century later, the word "virus" had until that point been associated with a darkly mysterious etiology: with miasmas, with poisons, with plagues. Rabies behaved as though it were a microbic contagion, and so Pasteur maintained absolute faith that it was one, even though he could neither culture it in broth nor observe it under the light microscope. The word "virus" conveyed his uncertainty of rabies' specific form and characteristics. It was not until 1898 that a "virus" was scientifically defined as a microbe that is invisible under the light microscope and can pass through a filter designed to trap bacteria; it was not until 1903 that it was experimentally demonstrated that the agent of rabies fit squarely within this category.

Despite the confounding invisibility of rabies, despite the fact that it seemed to violate the scientific principles of the day to do so (principles that Pasteur himself had played no small part in establishing), Pasteur persevered in his work on a vaccine. His intellectual flexibility in the face of unexpected results allowed him to conclude early on that trying to cultivate the agent of rabies using existing laboratory methods would be fruitless. Instead, he nimbly refocused his attention and that of his assistants on inducing immunity, in animals and eventually humans, to what would remain an obscure, intangible foe.

Bourrel's two rabid dogs were part of a surge of rabies cases in Paris during 1880, and so the Pasteur laboratory would have no trouble obtaining infectious material. They got it from kennels of the national veterinary school at Maisons-Alfort and from private veterinary offices around the city. Because rabies could not be cultured on a plate or in a vial, it had to be maintained in living tissue. In the 1880s, this meant within the corporeal cells of a living afflicted animal. The maintenance of rabid animals within the modest rooms and basement of the Pasteur laboratory was discomforting to the personnel. There was the ever-present risk of contracting rabies—either directly from the jaws of the animal, or at the bench top, where infected tissues and sharp instruments could combine to do harm. Meanwhile, the researchers were forced to weather the public fury of the antivivisectionists, who denounced their work as senseless torture of innocent creatures.

In order to create and test a vaccine against rabies, the Pasteur team first had to develop a strain of rabies that behaved more reliably than the natural infection. Early studies relied on the crude method of one animal biting another, followed by an anxious wait over weeks or longer to see whether infection had been transmitted. The Pasteurians developed a preference for inoculating their subjects, rather than exposing them to the competing risks that accompanied the bite from a rabid dog: trauma, sepsis, fright. However, this technique involved the dangerous collection of rabid saliva from a raging animal. Pasteur's son-in-law, René Vallery-Radot, recalled one such dramatic scene:

> "We absolutely have to inoculate the rabbits with this slaver," said M. Pasteur. Two helpers took a cord with a slip knot and threw it at the dog as one throws a lasso. The dog was caught and pulled to the edge of the cage. They seized it and tied its jaws together. The dog, choking with rage, its eyes bloodshot, and its body racked by furious spasms, was stretched out on a table while M. Pasteur, bending a finger's length away over this foaming head, aspirated a few drops of slaver through a thin tube. It was . . . at the sight of this awesome tête-à-tête that I saw M. Pasteur at his greatest.

Such exertions would soon prove unnecessary. Careful experiments showed that rabies could be as readily communicated with material from the affected animal's brain stem as with its saliva. "The seat of the rabic virus," wrote Pasteur, "is therefore not in the saliva only: the brain contains it to a degree of virulence at least equal to that of the saliva of rabid animals." Whether the Pasteurians were inoculating nervous tissue or slaver, uncertainty during a prolonged incubation period remained a problem, as not all animals would manifest signs of rabies following inoculation and the interval before onset of signs was still quite variable. "It is torture for the experimenter to be condemned to wait for months on end for the result of an experiment," lamented Pasteur.

The Pasteur team soon found it was able to improve the infection rate and shorten the incubation period by administering chloroform anesthesia to the recipient animal, trepanning a hole in its skull, and then inoculating the rabid nervous tissue directly onto the dura mater, the connective tissue that covers the brain. Pasteur, disturbed by the invasive nature of this procedure, initially resisted the widespread implementation. But he was soon reassured by the vigorous, happy appearance of his laboratory's first subject dog, only one day post-trepanation. The method was perhaps more stressful for the experimenters themselves, as remembered by Emile Roux's niece:

> [Roux], [Charles] Chamberland, and [Louis] Thuillier bent down around a table. A large dog was tied down on it, its muscles contracted and its fangs bared. . . . If the animal, despite all the precautions, had caused them to make a false move, if one of them had cut himself with his scalpel, and if a small piece of the rabid spinal cord had penetrated into the cut, there would have been weeks and weeks filled with the anguished question: will he or will he not come down with rabies? . . . They were no longer just "researchers" absorbed in the meticulous work of their laboratory; they were pioneers, adventurers of science.

Using the trepanation technique, Pasteur's assistants successfully transmitted rabies to the healthy animal in every case attempted. Signs of disease were apparent in the inoculated animal in less than two weeks—a significantly shorter time than with natural infection—and death concluded within a month. As canine rabies was thereby passed to a rabbit, and from one rabbit to another rabbit, and from that rabbit to still another rabbit, and so on in successive passages, the incubation period became reliably shorter. Once twenty-one passages had been made, brain to brain, one rabbit to another, the incubation period had decreased to eight days. Here it became fixed and constant, so that subsequent passages in rabbits produced no further change.

Shortened incubation period is associated with increased virulence. The enhanced virulence of rabies following intracranial serial passage in the rabbit was plainly observable when the virus was inoculated back into a canine host: dogs infected with the rabbit virus were even more catastrophically affected than those afflicted with natural strains. Through persistent repetition, Pasteur could now induce a consistently deadly version of the volatile virus at will. Even if he could not culture it in a tube like a bacterium, or coax it to shine in the eyepiece of his light microscope like a spore, the rabies virus was now finally under his control.

The next step would be attenuation: the deliberate weakening of the virus in order to induce immunity without causing disease. From the beginning, Pasteur had sought a strain with a sure and fast-acting immunity that could be applied *after* exposure to rabies had already occurred. Much more challenging to achieve than the already-established method of prophylaxis, the vaccine strain would race to establish immunity against a natural rabies infection as it made its murderous way from bite wound to brain. If the infection inhabited the brain before protective immunity had taken hold, the patient's death from rabies would be as certain as ever. But Pasteur hoped that an extremely robust yet attenuated rabies vaccine would provoke a sufficiently quick, vigorous immune response to interrupt the progress of the infection and spare the brain—indeed would drive the virus from the body altogether. The postexposure application of a paradoxically "weak strong" rabies vaccine strain would require innovation beyond Jennerian-Pasteurian vaccine principles. In fact, it would necessitate the creation of an entirely new branch of science: immunology.

Pasteur would create his highly immunogenic but determinately safe rabies vaccine strain through a two-stage process: a first stage that would carefully hone the virulence of the virus, and a second that would deliberately blunt it. The second stage, like the first stage, relied on the ingenious manipulation of postmortem nervous tissue from rabid animals. It also, like the first stage, was directly carried out by Pasteur's most trusted assistants: Chamberland, Thuillier, Adrien Loir, and, especially, Roux. Roux, in fact, probably invented the Pasteurian method of attenuating the most dangerous strains of rabies by aging the dissected spinal cords of rabbits that had succumbed to the virus in specialized flasks for desiccation (although Pasteur himself would take much of the credit for this). As they had done with so many other methodologies devised in the Pasteur laboratory, the research team perfected and verified this protocol through numerous repetitions. Soon they had gone on to demonstrate the powerful effectiveness of their attenuated-virulent strain as a vaccine—both as a prophylaxis against future exposure to rabies in dogs and, ultimately, as a postexposure immunization therapy that would prevent rabies in dogs already exposed to the deadly virus.

In September 1884, Pasteur received a letter from the emperor of Brazil, inquiring when a vaccine would become available for human victims of a rabid bite. He replied:

> Until now I have not dared to attempt anything on men, in spite of my own confidence in the result and the numerous opportunities afforded to me since my last reading at the Academy of Sciences. I fear too much that a failure might compromise the

> future, and I want first to accumulate successful cases in animals. Things in that direction are going very well indeed; I have already several examples of dogs made refractory after a rabietic bite. I take two dogs, cause them both to be bitten by a mad dog; I vaccinate the one and leave the other without any treatment: the latter dies and the first remains perfectly well.
>
> But even when I shall have multiplied examples of the prophylaxis of rabies in dogs, I think my hand will tremble when I go on to Mankind.

It was only six months later, in March 1885, that Pasteur wrote in a letter to his friend Jules Vercel, "I have not yet dared to treat human beings after bites from rabid dogs; but the time is not far off, and I am much inclined to begin with myself—inoculating myself with rabies, and then arresting the consequences; for I am beginning to feel very sure of my results."

Perhaps he continued to weigh the possibility, but Pasteur never did submit himself to this terrifying trial of his own vaccine. It was never necessary, as there were always unfortunate dog-bite victims whose physicians and families would offer them up for experimentation. Pasteur's notes indicate that he allowed himself to be persuaded more than once to administer a vaccine to humans already in the throes of hydrophobia. These patients received no benefit from vaccination, as the natural virus had already infected their brains at the time of inoculation. Pasteur drew the necessary scientific conclusions and then made sure that these false starts were never publicized. He continued to believe that under the right circumstances the rabies vaccine would succeed, and he was determined that the vaccine's public debut in humans would be nothing less than triumphant—providing the world with a broadly persuasive argument for the lifesaving potential of vaccines.

It was the destiny of Joseph Meister, a boy of nine, to provide Pasteur with a sufficiently compelling experimental case to deploy his fledgling vaccine. While walking alone to school on the outskirts of his small Alsatian village, Meister was viciously attacked by a grocer's dog. The animal knocked him to the ground and tore at his flesh while he cowered, holding his hands over his face in vain. By the time a nearby bricklayer reached the scene and fended off the dog with two iron bars, Meister had suffered fourteen penetrating wounds to his thighs, legs, and hand. Later that day, after cauterizing the bite wounds with carbolic acid, Meister's local physician sent the boy to distant Paris for consultation with the famous Louis Pasteur.

Pasteur proceeded cautiously. He was touched by his initial meeting with the stricken boy and his mother but did not agree to treat Meister until he had conferred with Alfred Vulpian, one of France's most respected physicians and a member of the government's official Commission on Rabies, and Jacques-Joseph Grancher, the head of the pediatric clinic at the Paris Children's Hospital. The two esteemed medical men agreed that experimental treatment with Pasteur's vaccine would offer Meister his best hope for survival given the extremely grave nature and severity of his wounds. Vulpian and Grancher provided not only an ethical sounding board for Pasteur but also very necessary practical assistance as he proceeded with his trial. Louis Pasteur had never been trained as a doctor, did not have a medical license, and so was prohibited from holding the syringe as it administered the first modern laboratory vaccine for humans, even though he himself had overseen every aspect of its creation.

Meister received his first injection immediately. "On 6 July, at eight o'clock in the evening, sixty hours after the bites of 4 July, and in the presence of Drs. Vulpian and Grancher, we inoculated into a fold of skin over young Meister's right hypochondrium half a Pravaz syringe of the spinal cord from a rabbit dead of rabies on 21 June; the cord had since then—that is, for fifteen days—been kept in a flask of dry air," recorded Pasteur in his laboratory notebook. The full, ten-day treatment would consist of thirteen inoculations, all delivering postmortem spinal tissue from a rabid rabbit. Each successive injection would contain a section of cord that had been exposed to air for a shorter time than the one before it, so that as the series proceeded, the vaccine would become less attenuated.

Throughout treatment, Meister and his mother were housed adjacent to Pasteur's lab at Collège Rollin. While Meister made himself happily at home among the laboratory chickens, rabbits, guinea pigs, and mice, Pasteur's dauntless confidence in the rabies vaccine wavered as the inoculations he dispensed became more and more virulent. "My dear children," began a letter from Mme Pasteur, "your father has had another bad night; he is dreading the last inoculations on the child. And yet there can be no drawing back now! The boy continues in perfect health."

On July 16, at eleven o'clock in the morning, Meister received his final inoculation. This preparation contained the most virulent tissue of all: rabid spinal cord from a dog that had been infected with a strain of rabies virus maximally strengthened by serial passage in the rabbit and harvested only one day prior to injection. Such a dangerous inoculation would provide a convincing test of Meister's immunity: a naive recipient of this shot would be expected to show signs of rabies within several days. Pasteur's son-in-law describes the fateful occasion as tense:

> Cured from his wounds, delighted with all he saw, gaily running about as if he had been in his own Alsatian farm, little Meister, whose blue eyes now showed neither fear nor shyness, merrily received the last inoculation; in the evening, after claiming a kiss from "dear Monsieur Pasteur," as he called him, he went to bed and slept peacefully. Pasteur spent a terrible night of insomnia; in those slow dark hours of night when all vision is distorted, Pasteur, losing sight of the accumulation of experiments which guaranteed his success, imagined that the little boy would die.

Shortly afterward, Pasteur left Paris for a much-needed rest and relied upon frequent updates from the physicians still monitoring Meister to reassure him of his successful treatment. On August 3, Pasteur wrote to his son from Arbois, "Very good news last night of the bitten lad. I am looking forward with great hopes to the time when I can draw a conclusion. It will be thirty-one days tomorrow since he was bitten."

As more weeks passed and Meister remained free from rabies symptoms, Pasteur began to share the news of his success with close associates. One of these, Léon Say, leaked the story to the *Journal des Débats,* and soon the world began tentatively cheering the news. After Pasteur returned to Paris in the early fall of 1885, he made a statement to the Académie des Sciences describing the treatment received by Meister. More than three months had passed since the child suffered his terrifying bite wounds, and still he appeared healthy. The details of the case were outlined for the academy. Dr. Vulpian rose first to respond:

> Hydrophobia, that dread disease against which all therapeutic measures had hitherto failed, has at last found a remedy. M. Pasteur, who has been preceded by no one on this path, has been led by a series of investigations unceasingly carried on for several years, to create a method of treatment by means of which the development of hydrophobia can *infallibly* be prevented in a patient recently bitten by a rabid dog. I say infallibly, because, after what I have seen at M. Pasteur's laboratory, I do not doubt the constant success of this treatment when it is put into full practice a few days only after a rabic bite.
>
> It is now necessary to see about organizing an installation for the treatment of hydrophobia by M. Pasteur's method. Every person bitten by a rabid dog must be given the opportunity of benefiting from this great discovery, which will seal the fame of our illustrious colleague and bring glory to our whole country.

Pasteur's modest laboratory at the École Normale was immediately transformed into a clinic and dispensary. People terrified of rabies arrived in droves to receive inoculations. By December, eighty courses of treatment had been completed or were in progress in Pasteur's bustling lab on the rue d'Ulm.

Every morning, Pasteur's assistant Eugène Viala meticulously prepared inoculations for the day's vaccinations. From rows of desiccating flasks, Viala selected and sectioned pieces of aged rabbit spinal cord. The pieces then were isolated in sterilized vials according to the number of days since postmortem harvest and suspended in a few drops of veal broth to create an inoculant. Pasteur supervised Viala's work closely and saw that, for every patient, an appropriately attenuated inoculation was prepared specifically for each given day of treatment.

At eleven o'clock, Pasteur's study was opened to patients. For each, the date and circumstances of the bite, along with the veterinarian's certificate, were entered into the register alongside the name of the victim. Pasteur stood attentively alongside the pediatrician Jacques-Joseph Grancher as he made each injection according to protocol. The patients and their families were free to ask questions of the celebrated scientist responsible for the vaccine, but these were often redirected to Grancher: Pasteur would never hesitate to gently remind his visitors that he was trained as a chemist, not as a physician. Pasteur's son-in-law recalls that "he had a kind word for every one, often substantial help for the very poor. The children interested him the most; whether severely bitten or frightened at the inoculation, he dried their tears and consoled them."

In December 1885, a telegram arrived at the rue d'Ulm announcing that four children from New Jersey, bitten by rabid dogs, were en route to Paris to receive Pasteur's now internationally famous cure. Money for their passage had been raised through a public subscription organized in the *New York Herald*. The published appeal, written by the well-known Newark physician Dr. William O'Gorman, read:

> I have such confidence in the preventive forces of inoculation by mitigated virus that were it my misfortune to be bitten by a rabid

> dog, I would board the first Atlantic steamer, go straight to Paris and, full of hope, place myself immediately in the hands of Pasteur. . . . If the parents be poor, I appeal to the medical profession and to the humane of all classes to help send these poor children where there is almost a certainty of prevention and cure. Let us prove to the world that we are intelligent enough to appreciate the advance of science and liberal and humane enough to help those who cannot help themselves.

Contributors to the subscription included Andrew Carnegie and the former secretary of state Frederick Frelinghuysen, along with neighbors and friends from the children's Newark neighborhood, whose nickels, dimes, and dollars within twenty-four hours had amassed to a thousand dollars. The four boys quickly embarked for Paris, accompanied by a doctor and by the mother of the youngest among them. That boy, only five years old, reportedly exclaimed upon experiencing the trifling sting of his first injection, "Is this all we have come such a long journey for?" As their treatment proceeded, the story was raptly followed by the New York press—whose articles were subsequently reprinted in newspapers across the nation. As much as 10 percent of the *Herald* was devoted to rabies while the children were under Pasteur's care. All of America waited breathlessly for news of the boys' cure.

When the healthy, vaccinated boys stepped off the boat from Paris several weeks later, they were celebrities in New York and across the nation. For months afterward, the four of them were trotted out in theaters and dime museums, from the Bowery in Manhattan to the heartland of America. For ten cents, the curious could witness with his or her own eyes the ongoing health and vigor of the treated boys, and even question them about their experience in Pasteur's laboratory. Around the United States and around the world, the media of the day continued to dwell on the particulars of Pasteur's rabies treatment as experienced by the four young Americans. Many newspapers also went out of their way to explain Pasteur's laboratory-based scientific research that gave rise to the vaccine.

According to the historian Bert Hansen, the popular sensation caused by the Newark boys receiving Pasteur's cure led to a profound change in the way Americans thought about science and medicine. "It reversed the assumption that older doctors and older medicines were better than new ones," explains Hansen. "It created a new expectation that medicine can and should change, that progress is to be expected, that the new advances would come from laboratory experiments on animals, and that specific injections would be a major tool of the new medicine." The public, led by journalists and public officials, now waited breathlessly for the arrival of new medical breakthroughs and greeted these with ready fanaticism. Some, in the decade or so following Pasteur's rabies vaccine, would prove to be worthy—like diphtheria antitoxin and diagnostic X-rays—while others would fall flat, such as Koch's tuberculin treatment for consumption. Meanwhile, the global medical establishment was forced to adapt to the popular view. In the French journal *Concours Médical,* one Dr. Jeanne editorialized in 1895:

> From the heights of our settled situations, we should no longer laugh at bacilli and culture media. Those who cultivate them already deserve our respect for the services that they have given mankind; for us, the old guard of the medical profession, they must also inspire salutary fear and a determination to be useful. We must march with the times. The coming century will see the blossoming of a new medicine: let us devote what is left of this century to studying it.
>
> Let us go back to school and prepare the ground for an evolution, if we are to avoid a revolution.

Before the four children had even begun the return voyage from Paris, enthusiastic groups of physicians in New York, Newark, and

St. Louis had initiated steps to bring Pasteur's cure to the United States. Pasteur made it known that he would welcome American scientists, along with those from all corners of the globe, to study his methods in his laboratory. By the year 1900, there would be at least six clinics devoted to administering rabies vaccines in the United States.

Back in Paris, having assembled enough cases to demonstrate a statistical difference in survival between those vaccinated and those not, Pasteur set his sights on creating an institution that could meet the growing demand for his rabies treatment, as well as provide a home for the ongoing scientific research that might lead to even more cures. Although fervently proud of his contribution to the glory of France, he wanted this establishment to remain independent of the government. On announcing a fund-raising campaign in 1887, Pasteur immediately began to receive donations from all around the world. From the editor of the *Herald*, to the tsar of Russia, to little Joseph Meister in Alsace, donors gave generously to the cause. But much was needed in order to endow Pasteur's grand vision. Around Paris, Pasteur became a philanthropic fixture, regularly appearing at charity balls, bazaars, and banquets—and in the drawing rooms of wealthy Parisians, discreetly soliciting financial contributions. His personal contribution of 100,000 francs made Pasteur himself one of the largest single donors to his own cause. On November 14, 1888, the Institut Pasteur was formally inaugurated.

The Institut Pasteur would serve as the flagship for the growing syndicate of Pasteur Institutes worldwide. According to its official statute, registered in 1887, the institute's purposes were "(1) the treatment of rabies according to the method developed by M. Pasteur; (2) the study of virulent and contagious diseases." Unofficially, the Institut Pasteur was intended to foster science that would not only protect human lives and livelihoods but also engender profitable applications to support the institute's self-perpetuation and growth. The modern buildings, erected according to Pasteur's specifications on an expansive property in the then-suburban Parisian plain of Grenelle, would house laboratories, kennels, libraries, and Pasteur's own comfortably appointed residence. Its opening ceremony was attended by the president of the French Republic; ambassadors from Turkey, Italy, and Brazil; the most esteemed French scientists of the day; and a robust international press corps, who would ensure that the triumphant opening remained prominent in newspapers worldwide for several days.

Even as Pasteur was seeing his doctrines grandly institutionalized, in Paris and around the globe, he was constantly under attack from scientific detractors. Foreign microbiologists, especially those in Germany and Italy, claimed that they could not reproduce his rabies vaccine results. Physicians at home and abroad insisted that the improvements in survival from rabies due to being vaccinated were insignificant. Scientific journalists, who had risen to prominence in Europe during the latter half of the nineteenth century because of the popularization of intellectual issues, for the most part supported Pasteur, but those who chose to make their careers questioning the contemporary scientific orthodoxy missed no opportunity to chip away at Pasteurian principles. Each time the vaccine failed to save a life, even if it was simply because the treatment was delivered too late in the course of disease, the case would occasion a whole new trial of Pasteur's methods in the dock of a skeptical press. Numerous publications gave column space to Pasteur's scientific rivals, further fanning controversy. Some writers emphasized alternatives to Pasteurian treatment. Others expressed a nostalgic view that traditional methods were better or even argued that Pasteur's vaccine was somehow derivative of historical therapies.

If Pasteur's contemporary popularity and eventual historical legacy did not suffer, it is because he never gave anyone else the last word about his research and its fruits. Each and every hostile article, whether published in an academic journal or in a popular magazine, would receive an aggressively didactic response from the man himself. To a Naples newspaper, Pasteur wrote about one of his rivals, "Dr. von

Frisch . . . has not succeeded, I am sorry to say. But I can counter his trials with positive results that will overthrow any negative facts he claims to have obtained." To his family, Pasteur remarked in frustration, "How difficult it is to obtain the triumph of truth! Opposition is a useful stimulant, but bad faith is such a pitiable thing. How is it that they are not struck with the results shown by statistics?"

Pasteur would die at home on September 28, 1895. His health, during the years leading up to his demise, had been undermined by a series of strokes, as well as by the confining fatigue of congestive heart failure. His last years were spent in somewhat diminished productivity at the institute, as described by Mme Pasteur in 1893: "Pasteur continues to be fairly well, but he must resign himself to put aside all work that is in any way strenuous. He takes much interest in the work of others. He still enjoys going to the Academies."

The "work of others" was Pasteur's principal source of pride in those final years, especially as those other scientists were frequently men he had trained himself at the École Normale Supérieure or who had learned their discipline at the Institut Pasteur. "Our only consolation, as we feel our own strength failing us, is to feel that we may help those who come after us to do more and to do better than ourselves, fixing their eyes as they can on the great horizons of which we only had a glimpse," pronounced Pasteur, with characteristic gallantry. Many of the Pasteurians would eventually be remembered for their own contributions to science and medicine—though acknowledgment of their individual achievements would not generally be realized until after Pasteur's day-to-day involvement in laboratory activities had decreased.

Emile Roux, Pasteur's closest collaborator during the creation of vaccines against fowl cholera and anthrax, and who had been so instrumental in devising an attenuation method for the rabies virus, would go on to develop serum therapy against diphtheria toxin. Élie Metchnikoff, a Russian biologist who had trained in Germany with

Koch, would help, during his time at the Institut Pasteur, to lay the scientific foundations of immunology by describing the mechanisms of cellular immunity. Albert Calmette, after establishing a Pasteur Institute in Saigon, would build on the antitoxin research of Roux and develop antivenom serum therapy for snakebites. Later, at a Pasteur Institute he had founded in Lille, Calmette would join Jules Guérin, another Pasteur disciple, in his work on tuberculosis; together they would identify the famous BCG (Bacillus Calmette-Guérin), a strain of bovine TB that functioned as a human vaccine. Alexandre Yersin, a Swiss physician who was working under Roux when the Institut Pasteur was inaugurated, went on to spend his most productive years in Indochina. When a plague outbreak threatened Hong Kong in 1894, Yersin quickly established a field laboratory in the afflicted city and within days had discovered the plague bacillus. He furthermore determined that the dead rats littering Hong Kong were the origin of the deadly epidemic and quickly developed and began production of a life-saving serum therapy against plague. Charles Nicolle, who met Pasteur only once, worked under Roux and Metchnikoff, then later at the Pasteur Institute in Tunis. He determined that typhus was spread by the human louse and that leishmaniasis was transmitted from dogs to humans by the bite of the sand fly. Jules Bordet, who worked in Metchnikoff's laboratory from 1894 to 1901, made great progress in the field of immunology, particularly concerning humoral immunity, and he discovered the bacillus responsible for whooping cough after creating a Pasteur Institute in Belgium. Together, these early Pasteurians would further the laboratory-based approach to medical problems favored by their master and would carry his doctrines linking science, medicine, and public health to all the corners of the earth.

Pasteur's remains were interred not in the Panthéon but instead, according to his family's wishes, in a specially appointed crypt beneath the Institut Pasteur. There, fifteen years later, his wife, Marie, would be laid to rest also. Mosaics depicting Pasteur's research triumphs watched over the tombs—and so did Joseph Meister, who, years after being the first to be vaccinated successfully against the horror of rabies, became the concierge of the institute. When the Nazis, on occupying Paris, attempted to visit the Pasteur crypt in 1940, Meister bravely refused to unlock the gate for them. Soon after this discouraging event, he took his own life.

Then, as now, Pasteurian science remained very much alive. Soldiers at the front in that war, on both sides of the battle, were protected from disease with Pasteurian vaccines, treated for illness with Pasteurian sero-therapies, and benefited from hygienic first aid and surgical techniques based on Pasteurian asepsis. As remains the case today, there were then still scientists ready to argue against Pasteurian principles—but history would take little note. Indeed, though many miraculous cures lay in the future, no figure in medicine since has ever enjoyed the heroic status conferred upon Louis Pasteur, conqueror of rabies.

Cover of a pulp horror novel, 1977.

6

THE ZOONOTIC CENTURY

It is impossible to overstate how utterly Louis Pasteur, during just two decades of work in the late nineteenth century, remade mankind's understanding of rabies. His great discovery did not just radically reduce the number of humans dying from hydrophobia in the West each year. Through his invention of a preventative rabies vaccine for dogs, he also significantly reduced the incidence of the disease in the animal most responsible for spreading it. Moreover, during the course of the twentieth century, Pasteur's treatment for humans was improved and refined. Growing the vaccine in duck embryos (and later in cell cultures), rather than in rabbits, simplified the process and standardized the product. Researchers eventually discovered that supplementing the postexposure vaccine with rabies immunoglobulin, derived from the blood plasma of already-vaccinated humans, would markedly improve the success rate. As death from rabies declined in the West, the disease came to exist in the public consciousness largely as an archaic holdover from an earlier age: seldom seen, nearly mythical. Not only did rabies cease to be a meaningful cause of death in industrialized countries; it became largely absent from the streets and lanes. No longer did rabies threaten to invade the home, to colonize the trusted creature sleeping at the hearth.

At the same time, though, even as the oldest animal infection receded as a public menace, the pioneering work of microbiologists was establishing that a whole host of other diseases were, in fact, linked to similar diseases in beasts. This includes, of course, the titan of them all, the bubonic plague. (The flea-borne pathogen that Alexandre Yersin isolated in 1894 was later renamed in his honor—*Yersinia pestis*—and in 2010, using skeletons from plague pits, it was proved beyond doubt that this pathogen had been the cause of the Black Death.) All of human history prior to the twentieth century had been haunted by rabies as an unforestallable and invariably fatal infection from animals. In the twentieth century, as rabies receded, it was replaced by a rapid succession of equally horrific and, in many cases, far more dangerous zoonoses.

The most fatal of these struck like a tsunami between 1917 and 1920, when some 50 to 100 million people worldwide—roughly 3 percent of the global population—died of a particularly nasty strain of influenza, the so-called Spanish flu. At the time, medicine believed influenza to be a uniquely human malady. But throughout 1918, during the height of the epidemic, reports trickled in about uncommon animal ailments that seemed to mirror the symptoms of flu. At a veterinary hospital of the French army, a shocking number of horses had been laid low with such a syndrome. In South Africa and Madagascar, it was primates, baboons, and monkeys felled by the hundreds; in northern Ontario it was moose, dead in the brush. But most ravaged of all was the Iowan pig. First noted at the Cedar Rapids Swine Show in the fall of 1918, flu-like symptoms spread over the succeeding months to literally millions of hogs. The veterinarian J. S. Koen, who tracked the disease while in the employ of the federal Bureau of Animal Industry, reported its toll in stark terms. "Sudden and severe onslaught. Patient very sick and distressed. . . . Muscular soreness, nervous and excitable. Congestion of eyes. Watery discharge from eyes and nostrils. . . . Temperatures usually high, many instances up to 108°F. Rapid loss of flesh, may lose as much as five pounds per day. Extreme physical weakness. Very rapid progress through herd. Lasts four or five days and patient begins to recover about the time death is expected." But recovery never arrived for the thousands of pigs, or roughly 1 to 2 percent of cases, that perished from the infection.

Just calling this pig malady a "flu" was enough to invite skepticism from scientists and hand-wringing from the pork industry, which worried that reports of a deadly human flu in pigs could turn the public stomach against its products. But Koen stood his ground. "I believe I have as much to support this diagnosis in pigs as the physicians have to support a similar diagnosis in man," he wrote, and went on:

> The similarity of the epidemic among people and the epizootic among pigs was so close, the reports so frequent, that an outbreak in the family would be followed immediately by an outbreak among the hogs, and vice versa, as to present a most striking coincidence if not suggesting a close relation between the two conditions. It looked like "flu," it presented the identical symptoms of "flu," it terminated like "flu," and until proved it was not "flu," I shall stand by that diagnosis.

It was, as the historian Alfred W. Crosby later wrote, "a peroration . . . worthy of Luther standing before the Emperor at Worms."

Over the course of the next twenty years, four scientists working on both sides of the Atlantic—Richard Shope, based in Princeton, New Jersey, at the laboratories of the Rockefeller Institute for Medical Research, and a team of three in the United Kingdom (Wilson Smith, Christopher Andrewes, and Patrick Laidlaw)—endeavored to prove Koen's intemperate assertion correct. In 1928, Shope was back in his home state of Iowa investigating so-called hog cholera (in fact, unrelated to the human virus of that name) when he started looking into Koen's theory. The following fall, when the swine infection hit the herds again in full force, Shope returned to collect tissue samples and mucus secretions, and by 1931 he and his mentor, Paul Lewis, had isolated what would later be confirmed as the swine flu virus.

Meanwhile, the three English scientists, also at work on the problem of flu, employed a domesticated animal that until then had stayed on the sidelines of the quandary: namely, the ferret. In the mid-1920s, Laidlaw had coauthored some pathbreaking research on distemper, a nasty respiratory illness in dogs that he and his collaborator demonstrated could easily be spread to ferrets. Now he and his collaborators on flu were attempting to pass the human strain of the disease to ferrets. In 1933 they succeeded, isolating a virus from diseased humans that created a comparable illness in twenty-six ferrets. Perhaps more important, they also demonstrated that Pfeiffer's bacillus, a bacterium that for decades had been suspected as the cause of influenza, had no effect at all.

That same year Shope, with whom the three were now working in excited collaboration, used his swine virus to produce flu in ferrets. And this was not just any flu; it was a horrific, Spanish-style flu, characterized by the same bloody pneumonia that had swept away millions in the global pandemic. By May 1935, in a lecture at St. John's College, Laidlaw was able to state confidently his opinion—one borne out by all subsequent discoveries—that "the virus of swine influenza is really the virus of the great pandemic of 1918 adapted to the pig and persisting in that species ever since."

This retrospective understanding of the 1918 pandemic was just the first such awakening in a century replete with terrifying zoonoses. U.S. soldiers returned from the Korean War with a sinister strain of hantavirus, a nasty hemorrhagic fever acquired from rats. In the late 1960s, Africans were falling ill from Lassa fever, also the handiwork of rats. The 1980s saw the emergence not just of AIDS but of the terrifying Ebola, also a disease from monkeys that, at its most acute, can induce terrible death involving bleeding from all bodily orifices. More recently, two strains of animal influenza—the onset of the avian flu and the robust return of the swine flu—killed thousands of people and prompted thousands more to hide for months behind surgical masks. Despite the slow ebb of rabies as a killer of men, the preceding century nevertheless supplied humans with countless reasons to eye their animal neighbors warily.

By the 1930s, rabies in America had largely subsided in humans, though their dogs were a different matter. This was especially the case in the South; a report from late in that decade put the infection rate among the dogs of Birmingham, Alabama, at 1 percent—a shocking number for a virus that kills all of its canine hosts. Though the canine vaccine was by then widely available, whites in the region vaccinated just 40 percent of their dogs, and African Americans only one in ten. The human death rate had been brought low but remained real: more than 250 deaths were logged in the American South during the 1930s, a per capita rate not dissimilar to the pre-Pasteur hysterias in eighteenth-century England.

That decade also saw the publication of a seminal novel, one of the most important of the twentieth century, whose denouement hinges on a spectacular demise in the jaws of rabies. *Their Eyes Were Watching God,* by Zora Neale Hurston, details the life and romantic entanglements of Janie Crawford, a black woman from a modest upbringing in western Florida. After being raised by her grandmother, a former slave, Janie marries a much older farmer; is seduced away from him by a soon-to-be politician and entrepreneur; then finally (after the death of this brutal and abusive second husband) falls in love with a much younger man named Vergible Woods, known to all as Tea Cake, whose fraught and tempestuous marriage to Janie consumes the second half of the book. The pair move first to Jacksonville and then to the Everglades, where they live happily in a shack, planting and picking beans by day and socializing at night, as Tea Cake entertains guests with his guitar and then takes their money (or vice versa) in dice games.

Their reverie is eventually shattered by the arrival of a hurricane, the ravages of which on their little shack prove to be the least of their troubles. Far more consequential, though neither is aware of this at the time, is the bite that Tea Cake receives—on his face, no less—from a shivering, furious dog they encounter while fleeing to safety. Perched improbably on the back of an almost wholly submerged cow, the dog "growled like a lion," with "stiff-standing hackles, stiff muscles, teeth uncovered as he lashed up his fury for the charge." Tea Cake wrestles with the dog and eventually drowns it, but not before it has clamped its slavering jaws upon his cheekbone.

Three weeks later, after the pair has fixed up a house in the Everglades and (they believe) resumed their life together, Tea Cake starts to complain of a headache. In the dead of night he wakes up in a "nightmarish struggle with an enemy that was at his throat." The next morning, offered a glass of water, Tea Cake gags on it, dashes the glass to the floor. "Dat water is somethin' wrong wid it," he exclaims. "It nelly choke me tuh death."

A doctor, summoned to Tea Cake's bedside, warns Janie that rabies is the likely diagnosis. The only thing she can do now is to leave him at the County Hospital, where the staff can "tie him down and look after him."

"But he don't like no hospital at all," Janie replies. "He'd think Ah wuz tired uh doin' fuh 'im, when God knows Ah ain't. Ah can't stand de idea us tyin' Tea Cake lak he wuz uh mad dawg."

"It almost amounts to dat," says the doctor, forebodingly. "He's got almost no chance to pull through and he's liable to bite somebody else, specially you, and then you'll be in the same fix he's in."

By the very next morning, Tea Cake has descended into a paranoid fury. Seized with suspicions of his wife, he returns from the outhouse and draws his pistol on her. She has prepared herself for this turn with a rifle. She plants a fatal shot in Tea Cake just as he lunges and bites her on the arm. She is forced to pry her slain husband's jaws from her own flesh.

Given the doctor's explicit warning, we are left to wonder: Has

Janie herself contracted rabies from Tea Cake's bite? Today's rabies experts believe that the virus is unlikely to be passed from human to human via biting. But as the Hurston scholar Robert Haas points out, that possibility was often emphasized by doctors in Hurston's time, and the doctor in her own book raises it. If Janie has been infected, we aren't given any sense that a similar madness is settling on her. And yet it's impossible to tell whether the window for infection has passed. The book both begins and ends at some unspecified time after the incident, when Janie returns to the town (Eatonville, Florida, Hurston's own hometown) where she and her second husband had lived and prospered. At her trial for Tea Cake's death—before, in a heavy irony, Janie is acquitted by an all-white jury that simply doesn't believe killing a black man to be a crime—the doctor testifies to finding her "all bit in the arm, sitting on the floor and petting Tea Cake's head." Probably we are supposed to conclude, somewhat reasonably, that he treated Janie then for her possible exposure.

The other lingering question, of course, is how Hurston came to use rabies as a plot point in the first place. Critics and biographers have found the choice somewhat bewildering. Robert Haas points out that Hurston's brother and first husband were both doctors, and her family had seven hounds while she was growing up; either of these might have lodged rabies in her brain, as it were.

But he also offers up a more intriguing and ultimately more plausible theory. At the beginning of 1936, as Hurston was writing her novel, a surprisingly popular movie about science played on screens all across the country. Its title? *The Story of Louis Pasteur.* Based on its gross of $665,000, Haas estimates that thirteen million Americans, or a full tenth of the population, would have seen the film—the entire last half of which is devoted to Pasteur's formulation of the rabies vaccine. In New York, where Hurston was living at the time, it ran for the whole month of February and also received reviews in the daily papers. It's only circumstantial evidence, to be sure; but we take a strange pleasure in the thought that the great Pasteur, while alleviating the terror of rabies in the streets, at the same time helped to inject a dollop of that terror into one of the century's great works of fiction.

It was also during the mid-1930s that the man who would revolutionize our ideas of the undead got his first taste of big-screen terror. The movie was *Werewolf of London*, and Richard Matheson was nine. "Somehow, I talked my mother into taking me," he recalled. "And when Henry Hull"—who in the film played a biologist, Wilfred Glendon, bitten by a strange animal while conducting research in Tibet—"changed into a werewolf, I freaked! Fell out of my seat and crawled up the aisle." The son of Norwegian immigrants, Matheson grew up in Brooklyn and excelled at science and music during high school, at the city's prestigious Brooklyn Tech. But after seeing combat in World War II and then graduating from the University of Missouri, he charted an entirely different course as a writer. He began with short stories in various genres—sci-fi, mystery, western—and then moved on to novels. To pay the bills, he worked days at the post office and later at an aircraft plant. At the time of his marriage, he had made only five hundred dollars from writing. "Those were very bad years," he later recalled, during which time his financial anxieties began to play into his fiction: "My theme in those years was of a man, isolated and alone, and assaulted on all sides by everything you could imagine."

In 1953, Matheson turned this trope into what is probably the most influential horror novel of the twentieth century. *I Am Legend* tracks the lonely existence of Robert Neville, who apparently stands as the only human survivor of a terrible virus that has killed off most of the people and turned the remainder into vampires. But these vampires are far from the becloaked aristocrats who haunted the dreams of the nineteenth century. They are insensate monsters who sleep all day in their lightless hovels and then roam at night in search of fresh blood. Immune to the virus, Neville becomes desperate in his loneliness, brought to the brink of self-destruction by the psychological ravages of his ceaseless routine: his home must be constantly fortified, and supplies replenished, in order to withstand the nightly onslaught. During the daylight hours he also drives around his city, a postapocalyptic Los Angeles, breezing down its abandoned avenues to stake as many vampires as he can find.

The slightest slip in this routine can lead to terrible consequences. One day, while making his rounds, Neville realizes that his watch has stopped; dusk is near, and he is at least an hour from home. At the intersection of Western and Compton he begins to see the vampires, rushing out of buildings as his station wagon passes. By the time he reaches his house, a mob of them await in front. He careers straight into the crowd, watching them fall like bowling pins, their pale, contorted faces crying out in agony. He heads past his house, and the remaining vampires make chase behind him, allowing him to dart around the block, park in front of his house, and dash to his door before they catch up.

We never learn the precise nature of the virus, though we do know that it afflicts dogs as well as people. Indeed, Matheson's description of a dog, dying in its agonies, gives us a sense that it bears an acute resemblance to rabies. Neville is amazed to see a live dog walking around, and so he begins to feed it. Soon, though, the dog succumbs: its expression begins to glaze, its tongue lolls out. Neville reaches for it, and its lips pull back in a threatening grimace. It begins to violently shake, with "guttural snarls bubbling in its throat." The dog dies a week later.

With *I Am Legend*'s evocation of a pandemic, and its intimations of nuclear devastation—World War III, we discover, has recently transpired, possibly helping to create the virus—the novel somehow managed to take the moribund vampire genre, still ruled even then by dour gothic Slavs in musty castles, and to reinvent it for the cold war era. One of Matheson's prime innovations was its setting: contemporary, suburban. But even more revolutionary was the nature of his "vampires." Far from the sophisticated loners of vampire literature to date, this was a mob of undead creatures whose threat lay not in their cunning but in their animal ferocity and, most important, their sheer numbers. "He was going out and staking vampires every day, finding them at the cold counter at Stop and Shop, laid out like lamb chops or something," Stephen King once said, in citing Matheson's book as a tremendous influence on his own work. "I realized then that horror didn't have to happen in a haunted castle; it could happen in the suburbs, on your street, maybe right next door."

Although *I Am Legend* calls its ghouls "vampires," the book actually was instrumental in jump-starting an entirely different genre: the zombie tale. The term "zombie" derives from Haitian religious belief and has been appropriated by American fiction authors since at least the late 1920s; Hollywood began making zombie movies in the 1930s. But in the mid-1960s, Matheson's story inspired George Romero, then a TV-ad director in Pittsburgh, to conceive of a more vital sort of zombie. In a short story that he eventually called "Anubis" (though never published), Romero "basically ripped off" Matheson's vision in describing a dystopian world where the dead have come back to life. Eventually, chafed by the constraints of television and unable to get funding for a feature, Romero and his friends decided to fund themselves in bringing his story to life. They pooled six hundred dollars apiece, from ten of them; a few of the other producers took roles in the film, the rest all pitched in as miscellaneous crew, and Romero directed. The result was a low-budget masterpiece called *Night of the Living Dead,* which grossed millions as a cult classic and also set the template for all zombie movies that would follow. Like the "vampires" in Matheson's story, Romero's zombie undead were not individual malefactors, not some garden-variety Draculas or wolf-men. They were an insatiable horde, eating their way through a society where all order has broken down. Zombies became synonymous with apocalypse.

The zombie-apocalypse genre has seen a particular resurgence in the twenty-first century. A 2004 remake of Romero's second zombie film, *Dawn of the Dead,* became a top grosser in both senses. Romero himself came off the bench to make the fourth film in his series, called *Land of the Dead.* A brilliant British spoof film, *Shaun of the Dead,* revolves around two London buddies whose instinct amid a zombie onslaught is to fight their way to their favorite pub. In books, Max Brooks's tongue-in-decaying-cheek primer, *The Zombie Survival Guide: Complete Protection from the Living Dead,* became a bestseller in 2003, and Brooks penned an even more successful follow-up novel called *World War Z;* meanwhile, *Pride and Prejudice and Zombies,* a version of the Jane Austen novel with flesh-eating "unmentionables" woven in at opportune moments throughout, shot up bestseller lists on both sides of the Atlantic. One graphic novel series, called *The Walking Dead,* has been turned into a popular TV show.

The *New York Times* has called zombies—in its Sunday Styles section, of all places—"the post-millennial ghoul of the moment." The question is, why? One theory is that the September 11 attacks took a peculiar psychic toll, leaving Anglophones with apocalypse on the brain. The sci-fi blog io9.com made a chart that purported to show zombies gained popularity during periods of social unrest. But its historical choices seemed fatally selective; for example, a long zombie dearth between 1943 and 1959 seems hard to square with this theory, given that Hiroshima and the rise of the cold war were two giant causes for apocalyptic musing if ever there were any. Another notion, which made the rounds during the 2008 presidential campaign in the United States, was that zombie booms correlated with Republican rule. Romero, after all, had reinvented the genre in the early days of Nixon, and then the Reagan administration ushered in a new wave that included *Re-animator* and *The Evil Dead.* In Democratic-leaning times, when (so the theory ran) popular rhetoric tends to demonize bloodsucking plutocrats, the Byronic vampire will find himself ascendant; in conservative periods, by contrast, the fear is heaped on mobs of shadowy masses—whether they be criminals or welfare recipients or Muslims—and so zombies naturally rise again to become the undead bugbear of choice. This theory, too, fails to convince: although

Obama's tenure has seen the rise of *Twilight*, the squishy tween vampire sensation, zombies have shown no signs of returning to their graves.

Before we can really parse the zombie wherefores, we need to recognize that there has not been just one zombie boom; there have been two concurrent ones, representing two very different visions of what a zombie can be. The first, and probably the most authentic to the Haitian origins of the term, is the slow zombie: the plodding, brainless variety, easily fought off one-on-one with a shovel to the head, or even a nice firm push to the torso. What makes the slow zombies dangerous is their sheer numbers and the relentlessness of their assault, day after day.* Slow zombies tend to also be more explicitly *undead,* in some cases even rising from graves, as in Romero's first film. Really, one should think of slow zombies as the true descendants of Arnod Paole, the ur-vampire observed by Johannes Flückinger in *Visum et repertum:* a creature devoid of cunning or fury, just a dead body walking the earth in a state of semi-decay.

But the fast zombie—well, that is a different beast entirely. These are more often than not the *infected* zombies, creatures of a fictional universe where a mysterious virus has descended on the population, spreading through bites and causing its human victims to become snarling, rapacious devourers of manflesh. Their means of murder is debased, to be sure, but their frenzy is not terribly far removed from the ancient *lyssa,* or wolfish rage, that spurred Heracles to slay his family or that swept Hector along to both glory and folly during the Trojan War. The fast zombie is a man (or woman) made into an insensate, murderous animal. The paragon of the fast zombie film, and almost certainly the best zombie film of the past decade, is *28 Days Later,* in which a virus called "rage" spreads through society. The film's

---

* It's this variety that best fits Chuck Klosterman's puckish notion of zombies as a metaphor for our high-tech modern life: "Continue the termination. Don't stop believing. Don't stop deleting. Return your voice mails and nod your agreements. This is the zombies' world, and we just live in it."

director, Danny Boyle, says he was specifically inspired by rabies, because it creates not just animal aggression in its victims but also an awareness, a mortal discomfort, as well as the physical horrors of hydrophobia. "We wanted the zombies to be bloodthirsty," Boyle says, "but completely full of fear themselves." In his desire to portray the agony of the zombie, he harks back to the legacy of Poe, carried down from Ovid, of horror tales forcing us to imagine the awful transformation affecting ourselves.

The most recent film adaptation of *I Am Legend*—the only to retain the original title—was another fast zombie affair, though not quite as effective as Boyle's lean masterpiece. As in the book, it's a disease that afflicts the monsters—they are pointedly *not* called vampires—though its particulars are roughly as convoluted as in the book; in the film's case, it's a modified version of measles, an engineered virus that cured cancer, on the plus side, but then unfortunately mutated to turn everybody into bloodthirsty ghouls. To see the transformation wrought in Neville's dog, though, makes the parallel to rabies even more explicit. In the film's telling, Neville (played by Will Smith) has a dog, Samantha, all along, and indeed Sam serves as his only companion for his daylight rovings and experimentations. When Sam gets bitten by an infected dog, and Neville is unsuccessful in saving her through serum, he sits on the floor of his lab embracing her while he waits to see what will transpire. Soon he gets his answer. Her pupils dilate; her teeth stretch into fangs; she begins to growl at her master. At the moment she lunges for his face, he regretfully converts his embrace into a choke hold; the most moving shot of the film is Neville's teary face as he strangles the life out of his one remaining friend.

To be clear: the fast zombie is not a rabid zombie, per se. These films are not in any sense *about* rabies, or about the fear of rabies; or, rather, if they are, it's only in the sense that the endorphins we feel on the treadmill are "about" the predator (not) nipping at our Nikes. A hundred years have passed since Americans have died from rabies in any meaningful number. And yet the basic trope of the fast zombie tale—the viral force that cuts out a soul, leaving a ravaging animal behind—has rabies woven deep into its DNA. Shielded from the disease, we nevertheless cannot wholly free ourselves from the fear.

The same year that Richard Matheson made the vampire more metaphorically rabid, the true rabidity of the vampire's animal sidekick became more widely appreciated. Late one summer morning in 1953, a seven-year-old boy on a Florida cattle ranch, in the shrubby pine flatwoods of Hillsborough County, near Tampa, was out searching for a lost ball when a depraved creature emerged from some bushes. It was a yellow bat—a species that eats nothing but insects—but today it seemed determined to make a meal of the boy. It latched on to his chest, holding fast even as he ran screaming to his mother. She knocked the bat to the ground, and the boy's father killed it. As he comforted his son, he remembered something he had read in a cattlemen's magazine, about how cows in Central America were catching rabies from vampire bat bites. He called up the local health authorities and insisted that they test the bat's carcass for the deadly disease.

For centuries, rabies had been known to scientists and citizens alike as a malady of the dog. Yes, the experts allowed, this disease could also manifest in other four-legged creatures: wolves and foxes, skunks and cattle. But when it came to bats—which harbor rabies far more frequently, and in a far more stable way, than any other species—science was shockingly slow to recognize the truth. Stretching back to Gonzalo Fernández de Oviedo y Valdés, observers have reported the bites of vampire bats to be "poisonous." Cattle in Central and South America are under constant assault from *vampiros,* and on rare occasions, often after a daytime attack, fully half of a cow herd would die from a terrible paralysis. Beginning in 1906, ranches in southern Brazil were decimated by a condition that came to be called *peste das cadeiras,* or the "plague of chairs"—so named because the hindquarters of the animals were being immobilized, forcing them into an unusual sitting pose. The cattle also salivated excessively, had difficulty swallowing.

As the paralysis slowly ascended, the animals became emaciated and finally died of respiratory failure. By 1908, more than four thousand head of cattle and one thousand horses had succumbed to this inexplicable disease.

In 1911, a São Paulo laboratory had a breakthrough: on the basis of characteristic spots, called Negri bodies, microscopically visible in brain tissue from fallen animals, it identified rabies as the culprit. And yet—even though vampire bats had commonly been seen biting affected herds in the unusual circumstance of broad daylight—the scientific community in Brazil was convinced that unseen dogs must be to blame. A major turning point came in 1916, when an epidemiological study pointed to vampire bats as the cause. That same year, rabies was positively diagnosed in a fruit bat. After hundreds of years of bat-borne rabies deaths in cattle, veterinarians and health officials slowly came to realize that *peste das cadeiras,* along with similarly ruinous livestock mortality events in Central and South America (*tumbi baba* in Paraguay, *rabia paresiante* in Argentina, *renguera* in Costa Rica, *derriengue* in Central America, *tronchado* in Mexico), was a devastation brought by aerial assault.

The first human deaths attributed to rabies spread by vampire bats occurred in Trinidad in 1929. Since dog rabies had been eliminated from the island in 1918, scientists were able to quickly and correctly assign the role of vector to the bat. In the three decades that followed, eighty-nine humans and thousands of domestic animals died from vampire bat rabies in Trinidad. But outside of Trinidad, it wasn't until the early 1950s that human deaths from vampire bat rabies were recognized. In 1951, a Mexican man, prior to succumbing to what would eventually be confirmed as rabies, told his doctors that four weeks prior he had suffered a penetrating bite while defending his children from an unusually aggressive vampire bat. Subsequent investigation by health officials revealed that in the man's home village of Platanito, four children had died of paralytic neurological disease since being bitten by the same bat.

By June 1953, when the life of the seven-year-old Florida boy was at stake after a daylight attack by a yellow bat, the prevalence of rabies in vampire bats was widely known. But North American bat species, of which the vast majority are insectivorous, were thought to be safe. Luckily, the boy's father was able to prevail upon health officials to perform the test. Within hours, W. R. Hoffert, a senior bacteriologist in the Tampa regional laboratory, saw the telltale Negri bodies in the bat's brain, a finding confirmed by further investigation at the Florida Board of Health in Jacksonville. The boy was given the postexposure vaccine, and he never came down with the disease.

News of this case prompted far more vigilant surveillance of rabies in American bats. By the end of 1965, infected bats had been identified in all states except Rhode Island, Alaska, and Hawaii; today, only Hawaii's bats are rabies-free. Bat bites are now the cause of nearly all human rabies infections in the United States, accounting for thirty-two out of thirty-three deaths from domestic exposure since 1990. Why is this so? Bat bites are so subtle that people can be infected without their being aware of it, especially in the night, when a bat bite is sometimes not even painful enough to wake a sleeping human. The U.S. Centers for Disease Control and Prevention recommend that anyone who awakens with a bat in his or her room seek out vaccination for rabies. Likewise, any unattended child or mentally incapacitated person found in the presence of a bat should be treated as though he or she were exposed. By foot, rabies may strike with snarling fury, but by air it arrives with the silent efficiency of an assassin.

Epidemics are always fraught with moral overtones, but never so much as in the case of AIDS, whose earliest populations of victims—gay men and intravenous drug users—were also among the most marginalized groups in Western culture. That the disease would also prove to be animal in origin, with all the cultural baggage that saddles such zoonotic infections, was unfortunate but unavoidable. What is more surprising is how many different animal suspects were considered before the right one was ID'd. In a letter to the *Lancet* in 1983, a Harvard researcher named Jane Teas published a brief letter fingering the pig, of all creatures: the first known Haitian cases (1978) of the syndrome, she pointed out, were followed soon afterward by a large-scale outbreak of African swine fever, a high-fatality infection in pigs that, like AIDS, affects the lymphatic system and weakens its hosts against other, opportunistic infections. Around the same time, the family pets, too, came into play. Media reports noted the similarity of AIDS to feline leukemia (which it resembles only somewhat) and canine distemper, or parvovirus (which it resembles hardly at all). Thus did the most terrifying illness of the twentieth century, a disease that changed an entire generation's consciousness about sexual behavior, begin with an uncommonly large menagerie of nonhuman perpetrators.

Speculation soon came to rest for good on the monkey. In late 1984, a research team at Harvard isolated a retrovirus from the blood of captive macaques that were suffering from AIDS-like symptoms. Unable to find a comparable virus in wild macaques, the scientists theorized that the macaques had caught the disease in captivity from a green monkey—whose wild brethren did, in fact, harbor a similar retrovirus. This "green monkey" theory catapulted into the public imagination, especially as the scientists began to outdo one another in their explicit theories for how the cross-infection transpired. A short June 1987 letter to the *Lancet* cited a 1973 anthropology paper reporting that the Congolese, in order to whip themselves into "intense sexual activity," injected monkey blood (from the corresponding gender of monkey, naturally) into their genital regions, as well as into their thighs and backs. The following month Abraham Karpas, a British AIDS researcher at the University of Cambridge, blew up this lurid just-so story into a full-page speculation in the *New Scientist*.

For more than a decade following that report, urban legends about the animal origins of AIDS abounded.* Much of what made AIDS so

* Today's scientists believe that AIDS most likely arose through the hunting of monkeys and apes for bush meat.

hysteria-inducing, of course, was the combination of its fatality rate (90 percent, in cases where infection with HIV progressed to the full-fledged syndrome) with the sexual mode of its transmission. It made a certain sick sense that such a beastly disease might be bestial in origin—might originate, that is, in bestiality. Anyone who grew up in late-1980s America, as we did, can attest to the variety and range of nonsense that circulated among adolescents of the time. But a similar set of stories swirled across most of the Western world. One AIDS researcher interviewed teens in her native Newfoundland and got responses like this:

> It originally came from Africa where they have the ritual practice of natives having sex with apes.

And this:

> The first one I heard was about a sailor whose ship stopped over in Africa and the sailor had intercourse with a baboon. The second story apparently happened in South America—Cuba, I think, or in Mexico. A man had intercourse with his sheep.

Meanwhile, in Scotland, a focus group convened by AIDS opinion researchers in 1990 yielded up this classic exchange of etiologic theorizing:

> **A:** I heard it was a guy had a thing with a gorilla.
>
> **B:** I heard it was a guy had sex with a bull.
>
> **C:** I heard it was a guy in Africa or something.
>
> **A:** It was just because of those black motherfuckers from abroad, man.
>
> **C:** Had sex with a gorilla or a monkey, something like that anyway. That's the way I saw it was the pakis that brought it here.

Note the way that each of these two groups, besides mentioning Africa, also fixate on the foreigners closer to hand. The Scots rail at the "pakis" (a slur that, having begun as shorthand for "Pakistanis," quickly evolved to encompass all Muslim immigrants to the United Kingdom), while the Newfies, for whom immigrants, let alone dark-skinned ones, are an uncommon sight, look nearer askance to Cuba or Mexico. One is reminded acutely of those dog-headed men whom the medieval cartographers sketched around the margins—but only in the lands far away from themselves.

Indeed, as one might expect, the residents of Latin America and Africa, including those most at risk for AIDS, see matters differently. In 1990, an AIDS researcher in Punta Gorda on the coast of Belize recorded this exchange between two women in a bar:

> **Woman bar owner:** That thing [AIDS] has been here since the beginning of time. It come from dogs, American women up there having sex with dogs, they catch it from them. Dogs. Here you don't let dogs in the house, they stay outside, cats too.
>
> **Woman farmer:** I heard it comes from those people in Africa who go to the forest and do things with monkeys. Those monkeys have it and give it to the people.
>
> **Woman bar owner** *(roaring with laughter):* A woman up there had sex with a dog and she gave it to her man. That's how it got started!

Even in Africa, undeniably the birthplace of the human virus, locals have an origin myth for AIDS that involves sex with a dog. In their version, which has circulated in Uganda, Kenya, Mali, and elsewhere in west Africa, it is in fact an African woman who has sex with the dog—but only because a white man paid her to do it. In Zimbabwe, the priest and anthropologist Alexander Rödlach traced this myth back to a 1991 story in Harare's *Sunday Mail* called "Inhuman Sex Acts: Women Arrested." Local police, the article claimed, "have confirmed the arrest of some women in Harare who were allegedly indulging in sex with a dog in exchange for money." The dog's owner, "believed to be a white man," would tape a video at each session, with the intention of selling the videos to pornographic markets "overseas." The paper quoted a supposed ex-boyfriend of one of the participants, who had confessed to him about her canine dalliances. What prompted her to come clean? He had confronted her about "why a venereal disease I had contracted had taken four months to heal." The former boyfriend is unnamed, of course; indeed, no sources are named anywhere in the article. AIDS is not mentioned, either.

More than a decade later, however, when Rödlach conducted wide-ranging interviews with Zimbabweans about the origins of AIDS, this dog story still came up frequently. Those who cited it usually did not believe that the infection had been in any way accidental. Instead, the white man had invented HIV, infected the dog with it, and then specifically recruited the black women in order to pass the disease on to them. Folk narratives of disease in Africa often blame some sort of "sorcery," and in this case it was a sorcery with a particularly twenty-first-century narrative, involving as it did an evil, virus-inventing scientist and an international pornography market. (In some African countries, the white man in the narrative has become a "European development expert.") And yet the core of the story is primal as well as universal, traded from Africa to Belize to Scotland to Newfoundland to the United States. Even in an era when science can illuminate the mysteries of disease at the finest molecular scale, we retain a deep-seated sense that an unnaturally virulent disease must have its origins in the most unnatural of couplings: the commingling of the human and the animal.

On May 6, 1994, when the Channel Tunnel began carrying rail traffic between the United Kingdom and France, the proximate terror in the minds of the British people was not about economic collapse or invading armies or even marauding tourists. It was about rabies. The disease had been eradicated entirely from Britain in 1902, and notwithstanding a few animal scares over the years, usually involving dogs brought in from mainland Europe, rabies had never again found a foothold on British soil. Just before the tunnel opened, one poll found that two-fifths of those who objected to the tunnel did so because it would make it "easy to bring rabies into the country"; in an earlier survey, carried out by a local paper in Folkestone, Kent, just near the tunnel's mouth, some 88 percent of respondents believed the Chunnel would render rabies "virtually unstoppable" or at least greatly increase its incidence. It is difficult to overstate just how large rabies loomed in the minds of the Kentish in particular. When interviewed by the Australian academic Eve Darian-Smith, an Anglican clergyman in Kent put it in the starkest possible terms. "The Channel Tunnel is a violation of our island integrity—a rape," he said. "Building it was a triumph of power and money over ordinary people and the English countryside. People think it might give us rabies in the same way as a rape victim might catch AIDS."

As not a few commentators suggested during the Chunnel dustup, the eradication of rabies at the century's beginning seemed if anything to have *increased* British terror of the disease in the subsequent decades. Rabies came to stand in for all manner of foreign ills; "the blessing of insularity," one member of Parliament remarked in 1990, "has long protected us against rabid dogs and dictators alike." And it did not help matters that an unscrupulous press had often preyed upon rabies fears in canny ways. This was particularly true during the mid-1970s, when a nasty outbreak of fox rabies in France made headlines across Britain. In the midst of the scare, Larry Lamb, editor of Rupert Murdoch's tabloid the *Sun,* bought for serialization a work of fiction called *Rabid,* which he retitled *Day of the Mad Dogs*. He had the first installment illustrated with the head of a rabid dog, foam running from its enormous jaws right down the page of the paper.

More shocking still, the paper produced a television commercial to promote the series, hiring as its canine stars the very same dogs that had appeared in the renowned horror film *The Omen*. "The commercial began sedately," Lamb later recalled. A middle-aged couple sits relaxing in the drawing room of an elegant country home. All of a sudden, the couple's two dogs pounce on them, slavering. "We showed close-ups of the dogs' foaming mouths"—achieved with shaving cream, Lamb said proudly—"and bloodied victims." This touched off a montage of horrors, including screaming babies and hunters out looking for mad dogs, before ending with a scene of the "expiring victim, sweating and moaning in hospital." The spot was so horrifying that by 11:00 p.m. on the very day it first ran, it was ordered off the air by broadcasting regulators.

Needless to say, *Day of the Mad Dogs,* both in serialization and then as a stand-alone novel, was a runaway success. Its terrible chain of events is set in motion by John and Paula, a young married couple who vacation in France soon after the death of their beloved dog. While staying in a villa just outside Cassis, the pair meets a bedraggled but affectionate stray, and Paula decides that they simply must bring it home. John is initially reluctant, but as he starts to come around, he finds that Paula, disinterested in sex ever since the previous dog died, notably warms to him again ("He began to wonder when he had last felt her nipple so erect"). Paula refuses to allow the new dog, which they name Asp, to get stuck in that nasty old British pet quarantine. Six months, she points out, is equivalent to "five years or more" for Asp; apparently, British dog years outpace American ones like the pound against the dollar. So the pair recruit John's rakish school chum Peter—with whom he once shared "a record unbroken stand for the first wicket against Lancing"—to smuggle Asp across the Channel in his yacht.

The outcome is as expected: once ensconced in their town of Abbotsfield, Asp goes mad. Soon the bodies, canine and human, begin to pile up. At the novel's end, with more than ten people (including

Paula) dead and a good fraction of Britain's pets exterminated wholesale, the townsfolk of Abbotsfield abduct John from his home and lock him in a dungeon with a rabid dog. Not until he is suffering through the final agonies of the disease do they haul him up from the prison, return him to his home, and set it ablaze with him inside.

The same year *Day of the Mad Dogs* appeared—1977—another pulp thriller about rabies, *The Rage,* hit British shelves.* Patriotic themes that in the former book merely hector the reader reappear in the latter to beat him about the head. In this story, it is Emma, ten-year-old daughter of the corrupt civil servant Lambert Diggery, who sneaks back an adorable dog she meets in the Ardennes, on a vacation where British declinism hangs morosely in the air. (The lass, after asking her father about the balance-of-trade deficit, sighs with innocent wisdom: "It's no way to run a country, is it?") Once smuggled into Mother England, the dog bites Emma's horse, which later rears up mad while the girl is riding it in a picturesque gymkhana. Next, the dog bites Emma herself, who, in the final throes of her own madness, will (in an echo of Tea Cake) chomp down on her own mother's neck. Finally the dog bites two foxes, the necessary narrative device by which to infect the hounds during a classic English hunt. As if this whole tableau were not isolationist enough, we learn along the way that Lambert Diggery is so besotted with the affections of Monique, a prostitute in Brussels—yes, he serves as a representative to the European Economic

---

* There is a curious American footnote to this peculiarly British outbreak of hysteria. For a few months in the fall of 1977, the horror writer Stephen King lived in England, during which time he penned the first draft of *Cujo,* America's most famous rabies-horror yarn. King has always maintained that the inspiration for the novel came from "reading a story in the paper in Portland, Maine, where this little kid was savaged by a Saint Bernard and killed." But it's hard to believe that he wasn't at least subconsciously influenced by the rabies-horror boom in Britain. He could hardly have been unaware of either book while in England, particularly because the latter book, *The Rage,* has an almost identical title to a novel of his own—*Rage*—that he had just published under his pen name, Richard Bachman.

Community, the predecessor to the EU—that he has been helping her sinister confederates move heroin onto British soil.

As in *Day of the Mad Dogs,* the imagined outbreak of *The Rage* could be easily controlled, were it not for a terribly implausible series of coincidences. One intrepid reporter locates the initial dog's corpse, but while driving to deliver the remains to his boss for testing, the young man crashes his car and is consumed, along with the evidence, in the ensuing fireball. Later the boss finds another rabid dog and locks it in his trunk, but when he is pulled over for reckless driving, the police open the trunk and allow the pooch to escape.

However ridiculous both novels may be, they do shed light on an essential problem of island eradication. Their plots hinge on the notion that authorities in both medicine and government, so sure that rabies cannot be a problem in Britain, will turn a blind eye to clear signs that the disease has returned. (Depressingly, this problem is far from fictional, as Chapter 8—about attempts to control a rabies outbreak on Bali, a previously rabies-free island—will make clear.)

So perhaps it is not surprising how keenly rabies figured, nearly two decades later, in the campaign against the Channel Tunnel. Opponents were little dissuaded by the argument that the length of the tunnel—some thirty-five miles long, with no source of food—would confound any four-footed migrants. These opponents were only partly mollified by the elaborate set of defenses that the tunnel's architects put in place after the outcry: security fences with animal-proof mesh, twenty-four-hour animal surveillance, and electrified barriers—"stun mats" was the more colorful term invoked—inside the tunnel. Soon before the tunnel opened, its PR handlers revealed to the media that a French fox had tested the defenses; their response was delicate but made clear that the unfortunate creature had not gotten far. Still, on the Chunnel's inauguration day in 1994, as Julian Barnes famously joked in *The New Yorker,* it was "as if lining up behind Mitterrand and the Queen as they cut the tricolor ribbons at Calais were packs of swivel-eyed dogs, fizzing foxes, and slavering squirrels, all waiting to jump on the first boxcar to Folkestone and sink their teeth into some Kentish flesh."

Fortunately, after nearly twenty years of operation, the rabies invasion of Britain has yet to materialize. The most recent rabid animal to be unwittingly imported was in 2008; it came not from France but from Sri Lanka, by air, and it was diagnosed while still in quarantine.

Sensationalism aside, westerners no longer have much reason to fear rabies as acutely as we do. Meanwhile, though, plenty of other zoonotic diseases—and their host species—are lining up to terrify us in the twenty-first century. From the monkeys, we have monkeypox, which more than ninety Americans contracted in 2003 after a batch of prairie dogs got infected at a pet store. Chikungunya and dengue fever, two more diseases that lurk in primate populations but spread via mosquito, have been expanding their range: in 2010, dengue was even found to be circulating in Miami. From the bats, we have the formidable Hendra and Nipah viruses, which cause encephalitis that kills human cases at rates upward of 50 percent. Nipah is perhaps the scariest of all, because it has already demonstrated its ability to spread from person to person; a survey of 122 human cases in Bangladesh found that 87 died from the disease, with more than half having been caused by human-to-human transmission.

Beyond these exotic new arrivals, of course, we have our annual bouts with the granddaddy of them all: influenza, whose yearly mutations wipe out thousands of people worldwide, with the threat of killing hundreds of times that when a particularly effective strain comes along. In 2009, it was the swine flu that snuffled back with a vengeance. Nearly three-quarters of a century after Patrick Laidlaw and Richard Shope identified the Spanish influenza as a disease of the pig, the H1N1 strain infected tens of millions of people, making it the first certified global pandemic since HIV/AIDS. The death toll was modest by pandemic standards but still significant, with more than fourteen thousand deaths confirmed and significantly more than that suspected.

Swine flu showed the incredible and abiding psychological power of animal origins in the cultural reception of disease. Once the flu's porcine origins had been revealed, there was little to stop the general public around the world from branding it a pig disease, despite all the caveats—no, you can't get it from eating pork—piled on at the urging of nervous governments. The standard naming convention for years had been to identify flu strains by country of origin, à la "Spanish influenza." But this proved to be even less tenable, politically, than the animal name: Mexican officials and commentators rose up in outrage at attempts to brand H1N1 the "Mexican flu," whereas the pigs could find no similarly eloquent advocates. "Swine flu" stuck.

The Muslim world took the swineness of the swine flu particularly hard. In Afghanistan, the nation's lone pig—Khanzir (Pig), which lives in the Kabul zoo—was placed unceremoniously into quarantine. Shops in the United Arab Emirates pulled all pork products from their shelves, and imports were suspended throughout the region. Tunisia went so far as to ban its citizens from carrying out the pilgrimage to Mecca, for fear that they might return with the disease. Swine flu did occasion some levity, at least, among newspaper cartoonists in Muslim countries, who used it as another excuse to tar their longtime adversaries (and mutual distrusters of pork), the Israelis. Qatar's *Al-Watan* newspaper, for example, ran a cartoon called "The Flu in Israel," in which one point of the six-pointed Jewish star formed a pig's head; less than a week later, it ran another sketch called "The Peace Process," in which an Arab is depicted as a surgeon, Israel as a flu-ridden pig. The same week in the UAE, *Al-Khalij*'s cartoon entitled "The Racism Flu" stuck a pig nose on the face of Israel's foreign minister. The following week, an Egyptian cleric, Sheikh Ali Osman, made this connection quite a bit more explicit when he declared that Jews were the source of all pigs and thus responsible for the outbreak.

Egypt, in fact, was the site of the most dramatic reaction to the pandemic: the wholesale slaughter of the country's 300,000 pigs, despite the lack (at the time of the edict) of a single documented case in either pigs or humans. These pigs were the treasured property of Egypt's Coptic Christian minority; in Cairo, a devoted population of Coptic trash collectors, called the *zabaleen,* had for decades employed thousands of scavenging pigs in disposing of the city's waste. But relations between Muslims and Christians have for decades been tense, sometimes erupting into outright violence, and it was hard not to see this preemptive move by the government as an act of prejudice carried out via the law.

A cynical view of the government's motives seemed especially justified once the methods of the massive pig cull became known. Officials had promised that the animals would be humanely slaughtered, with throats cut, and the meat preserved; but an Egyptian newspaper, after following a truck of seized pigs, found that their treatment was nothing short of barbaric. Video footage shows workers using a front-end loader to fill an enormous dump truck with screaming, squirming pigs, piled atop one another. Then the truck deposits the pigs at a vast burial ground, where they are killed slowly—"covered with chemical products and left for thirty or forty minutes until they are dead," one worker tells the camera—and then unceremoniously covered with quicklime. Other amateur footage showed pigs brained in the streets with metal poles, piglets stabbed to death.

It was a scene that could have played out in the nineteenth-century streets of London or Paris, except with a different animal as victim. For all the ways that scientific advancement during the past century or so has served to beat back superstition, it's worth reflecting that our zoonotic sleuthing in particular—our establishment, beginning with *Y. pestis* in 1894, that most of our myriad afflictions have an origin in animals—has been something of a Pandora's box. For centuries, after all, we suffered through waves of flu without having to blame some different creature for each set of shivers. Today, though, every new strain will expose some species to the sort of TV coverage normally reserved (at least where animals are concerned) for shark attacks alone. Four thousand years after the Laws of Eshnunna, and more than a century after Pasteur slew rabies, acquiring a disease from animals still shocks us, and, as the pigs of Cairo discovered, it can still drive humans to hysterical violence.

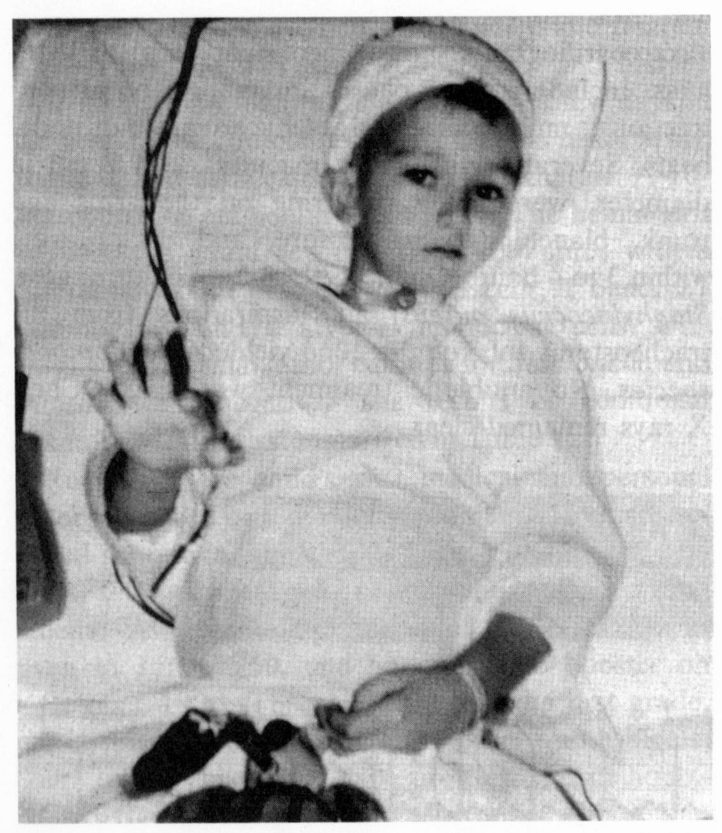

Photograph of Matthew Winkler, a six-year-old boy who survived rabies in 1970. Dr. Rodney Willoughby taped this photo to the wall of Jeanna Giese's hospital room in 2004.

7

THE SURVIVORS

Dr. Rodney Willoughby, a specialist in pediatric infectious disease at the Children's Hospital of Wisconsin, in Milwaukee, was dubious when he heard that a possible case of rabies was being transferred to his care. "I was skeptical she had rabies," he recalls. "Because that never happens."

It was October 2004. The patient was a high-school athlete, a fifteen-year-old girl who was suffering from fatigue, vomiting, vision disturbances, confusion, and loss of coordination. Willoughby considered some other brain infection or various autoimmune diseases as more likely causes for her condition. But he made sure that the samples necessary to rule out rabies were collected and sent to the Centers for Disease Control and Prevention (CDC) in Atlanta within hours of the girl's arrival. In the meantime, she was put in strict isolation to protect hospital personnel from possible exposure. Her condition quickly deteriorated; she began to salivate excessively and developed an involuntary jerking in her left arm. Soon, Willoughby had to sedate her and insert a breathing tube. As hours ticked by, he began to prepare for the possibility of a positive test result.

The girl's name was Jeanna Giese, and her troubles had begun a month beforehand, during a Sunday Mass at St. Patrick's Church in her hometown of Fond du Lac, Wisconsin. As she sat beside her mother, Giese observed the small silhouette of a silver-haired bat flitting against the sanctuary's tall stained-glass windows. When the bat fluttered down toward the back of the room, barely above the heads of the worshipping congregation, an attending usher batted the creature to the ground. Giese decided she would take it outside. With her mother's permission, she slipped quietly from her seat and walked back to where the bat lay prone. As she picked the bat up by the tips of its wings, it shrieked, but still she continued with it toward the door. Just as she nudged her way out into the open air, the bat reared its head around and bit its Good Samaritan on her left index finger.

Later, Giese showed the tiny wound to her mother, who ensured that it was thoroughly cleaned. No one in the family thought to seek postexposure treatment for rabies. But after symptoms set in four weeks later and Giese was admitted to a local hospital, her mother mentioned the bat bite to the pediatrician. Arrangements were immediately made for Giese's transfer to the Children's Hospital and into Dr. Willoughby's care.

Like the vast majority of American physicians, Willoughby had never seen a case of clinical rabies before. He telephoned the CDC to ask if there was any treatment for rabies somewhere in the research pipeline—some promising new therapy, perhaps, that had been attempted in a case or two but not yet published in any medical journal. The CDC could offer no such hope. Not one person had ever been shown to survive rabies without receiving at least partial vaccination against it prior to the onset of symptoms. All the treatments tried to date had failed. No consensus existed for what therapy should be attempted next. Aside from palliative care, standard practice was to use intensive therapy but in a purely reactive way, trying to control the dangerous complications of rabies as they arose. But this had never saved a single patient in Giese's predicament.

Willoughby attacked the problem with quick but deliberate reading. With less than a day to formulate a plan, he started out by searching for any recent papers that hinted at a possible treatment. None turned up. "I did a couple hours of diligence and figured out that nothing was new," he recalls.

So he decided he would use his limited time to review the basic neuroscience of rabies. His understanding—though the science is still far from settled on this subject—was that rabies did not cause inflammation in the brain, nor did it destroy the brain's slow-growing, densely networked cells. Instead, it seemed to interfere with how they communicated with one another, ultimately disabling the brain from performing crucial functions such as controlling cardiovascular activities and breathing.

Willoughby was struck with a novel idea for how to assist a patient through a rabies infection. The solution, he says, looking back, "was hiding in plain sight." He sat down at his computer and searched the scientific literature for the terms "rabies neurotransmitters" and "rabies neuroprotection" and then quickly tried to absorb the fifty or so papers his query returned. As he read on, he began to permit himself to hope that even if Giese was confirmed to have rabies, there might be a way to help her survive. "With a little more reading," he says, "it seemed to me like there was a real opportunity."

Willoughby had started thinking about becoming a physician when he was still in high school. His mother's father was a doctor, and Willoughby liked science, so it seemed a natural fit. He picked up the prerequisite courses as a Princeton undergraduate while still considering other possibilities; when none proved compelling, he enrolled in medical school at Johns Hopkins.

He most certainly did not become a doctor because of any burning desire to solve the human rabies problem. Not that he was unaware of the dreadful nature of the disease. During much of his childhood, Willoughby's large Catholic family lived in Peru, where his father worked for an American oil company. There, his younger sister was bitten by a guard dog that was defending the home of a family friend. The bite itself was not terribly serious, and if the dog had been observed to remain in good health over the next week or two, no further action might have been necessary. However, just after this incident, in the course of a burglary, someone threw poisoned meat over the broken-glass-topped concrete wall that surrounded the friend's property, killing the dog. Given the prevalence of canine rabies in Peru at the time, the Willoughby family did the prudent thing and started the girl on Pasteur's vaccine.

Willoughby himself would often accompany his sister to the clinic for her inoculations. It was clear that those fourteen shots, delivered into the sensitive muscles of her abdominal wall, were tremendously painful. But the injections were made much more frightening by the brutal manner of the German nurse who dispensed them. "Frau Nurse would tell her to toughen up, and then would slam the shot into her belly," he says. "The nurse was scarier than the shots were."

By the time Willoughby graduated from Johns Hopkins in 1977, human rabies had become vanishingly rare in the United States. "For the boards," he recalls, "you only needed to know one thing about rabies: it was 100 percent fatal." Willoughby committed this fact to memory, passed his boards, and didn't think much about the disease again for many years—even as he continued his training, first at the University of California at San Diego and then back at Johns Hopkins. "It's so rare in this country, only a few cases per year. So I figured I'd go pretty much forever without seeing one."

Willoughby would become a specialist in pediatric infectious disease, with a strong emphasis on clinical research. His work would center on diseases with importance in the developing world, such as rotavirus (a common and often fatal diarrheal infection in children) and cerebral palsy (which sometimes can be triggered by brain infection in young children). Along the way, his training exposed him to many talented clinicians and researchers. He was particularly impressed by Richard Moxon, now chair of pediatrics at Oxford, for the way he engaged in collaborative, open scientific discourse—to the point of being willing to share laboriously obtained DNA extracts from his laboratory with rival researchers. "That kind of openness to move the field forward, even if it doesn't benefit you personally, has always been inspirational," Willoughby says.

He had been practicing at the Children's Hospital of Wisconsin for only five months when Jeanna Giese came under his care. The night she arrived was just his second night on call. Treatment of Wisconsin's first human rabies patient in several years would turn out to be a great way to get to know his new colleagues and to reach out across the pediatric disciplines. With the help of his new boss, Michael "Joe" Chusid, Willoughby assembled a diverse team of talented clinicians. There were two neurologists, two criticalists, another infectious disease person, and an anesthesiologist—"a bunch of smart people," says Willoughby, each bringing a different but relevant area of expertise to his fast-moving conundrum.

At 4:30 p.m. on Giese's second day of hospitalization in Milwaukee, her test results came back from the CDC laboratory. She was positive for rabies, based on the presence of rabies antibodies in her blood and cerebrospinal fluid. None of the rabies virus itself could be recovered from her tissues, but based on her history and clinical signs, and in the absence of another likely cause for her symptoms, the positive antibody test was clinically adequate proof of rabies infection. An hour later, at 5:30, her physicians met at the hospital to discuss her treatment.

Willoughby brought to the meeting his idea for a new rabies treatment. He had developed it on the basis of two published assertions about the disease. The first was that rabies seemed—though this is somewhat controversial—to kill patients without causing any significant damage to their neuronal cells. The second was that the immune system does mount a response to rabies that could, in principle, fight off the infection. Willoughby had come to subscribe to the theory that rabies was a disease primarily affecting neurotransmission, or the electrochemical communication that takes place between the cells in the central nervous system. By disrupting signal transmission through the brain, so the theory went, rabies interrupted its ability to orchestrate such essential functions as breathing, blood pressure, and cardiac rhythm. These key roles are performed by what is called the autonomic nervous system—the unconscious, primitive seat of the brain. It is by disrupting the autonomic nervous system that rabies kills the patient, often through circulatory collapse or simple suffocation.

On Willoughby's theory, the battle against rabies was primarily a battle for time. Rabies wasn't killing the brain directly, but it was directing the brain to kill the body before the body had time to fight it off. Willoughby put a question to his colleagues at the Children's Hospital of Wisconsin: What if they induced a coma in Giese? By suppressing her brain activity, and by controlling her respiration and circulation—the functions of that autonomic nervous system—they would try to give her immune system the time it needed to mount its own response.

He gave his colleagues an opportunity to raise objections. "I set it up so that any of them could blackball it," Willoughby recalls. "If we had one blackball, then we wouldn't do it—because it was such a simple idea it had to be wrong. It was just too obvious. Someone had to have tried it before. So if anybody could see a reason why it would clearly cause harm, they could object and we would drop the plan. Instead, thirty minutes later, we didn't have an objection to it."

Later that evening, Willoughby met with Giese's shaken parents, Ann and John, to inform them of the test results and to discuss their daughter's grim prognosis. "We brought the parents in and gave them the bad news," recalls Willoughby. "The Giese parents, I think, especially John, still really didn't fully understand that this was irrevocable." He gave them three treatment options for their daughter: hospice care, which would allow their daughter a comfortable death at home; the standard critical care regimen, which so far had never been successful in saving an unvaccinated victim of rabies; or the experimental plan that Willoughby and his colleagues had laid out. The Gieses chose the third option without hesitation. They pointed out to Willoughby that even if their daughter didn't survive her infection, the knowledge gained might help some future child with rabies. But even as they said this, their hopes were fastening themselves securely to the idea of a miracle. (John Giese would later tell a reporter from the *Milwaukee Journal Sentinel* about the desperate optimism that helped him through this terrible moment. "Somebody has to be the first person to walk away from this," he recalls thinking. "Jeanna's going to be it.")

In the tense days that followed, the girl lay motionless in a hospital room, animated only by monitors and by the rhythmic whoosh of the mechanical ventilator. An infusion of ketamine, a dissociative anesthetic, maintained her state of unconsciousness. Willoughby chose ketamine for a particular reason: not only would it keep the patient in a state of coma, but it had been shown in a 1992 study on rats to have an antiviral effect against rabies. The effect of the ketamine was broadened by the addition of amantadine, an antiviral with a similar molecular mechanism of action but with an affinity for a different part of the brain. Midazolam, a sedative similar to Valium, was administered to smooth out the effects of the ketamine and to help maintain unconsciousness; this was supplemented occasionally with barbiturates, to keep the girl perfectly calm. On the second day, under counsel from the CDC, Willoughby added ribavirin, a broad-spectrum antiviral agent often used in treating hepatitis C. Nothing remotely resembling this regime, with its high-stakes induction of coma, had ever been administered to a rabies patient before. Tension pervaded the ward of the Children's Hospital of Wisconsin where Jeanna Giese slept unperturbed, busy nurses hovering above her.

Onto the wall above Giese in the pediatric ICU, Willoughby had tacked up a blurry black-and-white photograph. It showed the bright gaze of a six-year-old boy in another hospital bed, far from Milwaukee in both space and time. The boy's name was Matthew Winkler, and the photograph had been published with the report, in 1972, announcing his own recovery from rabies—the first scientifically supported case of survivorship ever published.

At 10:00 p.m. on the evening of October 10, 1970, six-year-old Matthew Winkler's sleep had been interrupted by a terrible pain in his left thumb. The boy awoke to find a brown bat fiercely clinging to his digit with its tiny jaws. The resulting clamor startled awake the entire Willshire, Ohio, farmhouse, bringing Winkler's father quickly to his bedside. The bat was wrenched free from Winkler's thumb, leaving two bleeding puncture wounds that the family cleaned thoroughly. The next day, the Winklers sent the bat off to the Ohio Department of Health, which identified rabies lesions in a cut section of its brain. The test results were reported back on October 14, and that same day Winkler's family physician initiated a fourteen-day course of duck-embryo vaccine. He did not, however, use immunoglobulin therapy—which by then was a common supplement to vaccination, providing a local immune response against the virus before the effects of the vaccine kick in. On October 30, two days after completing his inoculations, the boy began to complain to his parents of neck pain. Fever, loss of appetite, vomiting, and dizziness followed over the next few days, despite several doses of oral tetracycline initiated by the family doctor. Winkler was referred to pediatricians in Lima, Ohio, who admitted him to St. Rita's Hospital on November 4.

Over the next few days, Winkler's condition deteriorated precipitously. The normally studious and well-behaved first grader became uncoordinated, obstinate, unable to walk or write, then altogether mute. The left side of his body was markedly weak, and his bitten thumb tightened into a stiff flexion across his palm. Increased pressure in his skull necessitated the placement of a drainage catheter in the lateral ventricle of his brain. He developed frightening cardiac irregularities, as well as respiratory distress that could be relieved only with a tracheotomy and oxygen supplementation. Small seizures afflicted the left side of his body, and a rash appeared on his arms and torso. Winkler had slipped into a coma, however, and so was now mercifully unaware of the violent ordeal his body was undergoing.

Although no virus was isolated from Winkler's skin or saliva, or even on his brain biopsy, abundant rabies antibody was present in his blood serum—much more than would be expected in response to vaccination alone. Antibodies were also present in his cerebrospinal fluid, which is expected to contain antibody only in the presence of natural infection. Tests for alternative diagnoses, infectious and noninfectious, were all negative. A diagnosis of rabies was thus established, and hope for the boy's survival seemed bleak.

After days spent motionless in a coma, Winkler gradually began to show signs of improvement. First, he became able to sit up with assistance. By November 30 he was sitting up on his own and making squeaking sounds in an effort to speak. More improvements followed rapidly. On December 1, he said his first recognizable word, and by December 7 he could take a few steps on his own, although his left side was still notably impaired. After weeks of physical and speech therapy, Winkler's doctors declared him normal in both voice and intellect. He was discharged from the hospital on January 21, 1971—his seventh birthday. At a recheck in May, he was found to have no lingering neurological abnormalities.

In their 1972 report in the *Annals of Internal Medicine,* Winkler's clinicians—led by Dr. Michael A. Hattwick—tentatively attributed the boy's survival to one of three possible factors. The first was the vaccinations received prior to onset of clinical infection, though the authors noted that no prior victims of vaccine failure were known to have survived. The second was the possibility that Winkler had been infected with a relatively low-virulence strain of bat rabies—though the authors also acknowledged that a test performed at the Ohio Department of Health indicated a high degree of virulence in the infecting strain. The third was the advanced critical care measures employed at St. Rita's Hospital, including the intraventricular catheter, tracheostomy tube, antiseizure medications, and intensive nursing care. "Since no specific antiviral agent is known to be effective once symptoms have developed," the report concluded, "the treatment of clinical rabies must rely on aggressive supportive care. We now know that such care can cure."

Another, similar case of purported survivorship took place two years after Winkler's. On August 8, 1972, a forty-five-year-old Argentinean woman was bitten by her suddenly furious dog; within a few days, the dog succumbed to its illness. At first, her doctor treated the deep wounds in her arm with cleaning, suturing, and a dose of tetanus antitoxin. She did eventually begin postexposure vaccine treatment for rabies, ten days after the attack. But less than two weeks later, before she had even completed the fourteen-day course, she began to feel tingling in her left arm. On September 8, she was admitted to the hospital with headaches and depression. She was found to be feverish and weak, with neuromuscular spasms, particularly on her left side. A clinical diagnosis of rabies was made, confirmed by positive antibody titers in her blood and cerebrospinal fluid. But over the next few months, despite setbacks, her condition generally improved under intensive care. By September 1973, her doctors described her recovery as "nearly complete."

Since then, three additional "partial" recoveries from rabies in vaccinated patients have been described in medical case reports. One was a New York laboratory worker who inhaled rabies virus in the course of vaccine research. The second, a nine-year-old boy in Mexico, was bitten on the forehead and face by a dog that had already attacked twenty-five other dogs in the neighborhood. The third case, in India, resembled the second: a six-year-old girl bitten by a street dog. But in these three cases, the patients wound up with permanent handicaps after their infections, ranging from blindness and quadriplegia to severe brain damage.

All five survivors documented between 1972 and 2002 shared one important characteristic at the moment of developing their first symptoms: they had all received at least part of a course of vaccine against rabies. For each case like these, though, many more would perish despite having received some treatment before developing signs of illness. And for those who had never received vaccine when their illness set in, there was still no precedent of survival.

After Jeanna Giese had spent seven days in a coma, samples were taken of her blood and cerebrospinal fluid, which demonstrated a marked increase in the number of rabies virus antibodies compared with samples obtained on the first day of hospitalization. Her body was on the attack, striking back at the viral invasion. Giese's immune system had mounted a robust defense against the rabies virus and delivered it to the embattled central nervous system. With this welcome piece of news, her doctors began gradually to withdraw the anesthetics. The girl's return to consciousness was observed anxiously by Willoughby, who could not be sure what to expect. The medical literature had described survivorship among unvaccinated animals, he wryly notes—but in animal studies, "every time you get a survivor, you euthanize it."

Although the electroencephalographic findings improved after the withdrawal of ketamine, the only immediate change on Giese's physical exam was that her pupils became responsive to light. No other reflexes were apparent. Her limbs lay flaccid on the bed. Willoughby worried silently to himself. "Oh God, I created a lock-in," he thought—meaning someone who is conscious but unable to communicate or respond in any physical way. "It's, like, the worst thing you can do."

The idea that Giese might survive rabies only to be left severely disabled was a constant source of worry during the days and weeks that followed. But her steady, slow improvements kept Willoughby's worst-case scenarios at bay. Three days after the anesthesia was withdrawn, Giese's lower leg resumed kicking in response to the reflex hammer. Two days later, she regained eye movement. In two more days, she was raising her eyebrows in response to speech; then, a few days after that, she began to wiggle her toes, and to squeeze people's hands in response to commands. "Every day was something new, and it was just miraculous," recalls Willoughby with a slow shake of his head.

Giese was clearly responding to her environment, but doctor and family both craved more definitive evidence of her return. At that point, to test the girl's ability to recognize a familiar face, Dr. Willoughby and Ann Giese removed their protective face shields and stood side by side next to Jeanna's bed. Her eyes, held open by Willoughby, flickered between them briefly and then fixed on her mother. Clearly she was in there, after all.

From there, Giese's recovery took weeks of incremental improvement. She had to regain her alertness and her attention span, as well as her ability to communicate her thoughts and feelings. Only very gradually did she regain governance of her five-foot-ten frame: gesture, movement, expression, swallowing, and speech all had to be relearned. After a total of one month in medical isolation, Giese was transferred for intensive inpatient rehabilitation that lasted several more weeks.

On January 1, 2005, Giese finally left the hospital for her home in Fond du Lac, triumphantly crossing the threshold in a wheelchair pushed by her father, accompanied by her mother and three brothers. In the local TV news footage of her release, the towering teenage athlete appears diminished—a slumped and childlike figure clutching a floppy yellow stuffed dog in her lap. Facing her were nearly two years of intensive physiotherapy, during which she had to relearn all the basic skills of being human. She had to learn to crawl, then to stand, then to walk. In a video made by her doctors several weeks after her release, her obstacles were vividly apparent: despite an engaged, giggling demeanor, Giese is shown to be struggling to enunciate simple words and to be coping with limbs, particularly her left arm, that seem to have a mind of their own—spontaneously jerking, dancing, saluting.

But by the time a second video was made, a little more than a year later, Giese's dedication to her rehabilitation had clearly paid off. She already appeared much more physically self-possessed; only a subtle slurring and a rare stumble in her speech remained apparent. While her gait had not returned to the easy athletic lope of a three-sport student athlete, at both the walk and the run she appeared comfortable, if not quite fully coordinated. And her improvements have continued over the years. In the spring of 2011, Giese graduated from Lakeland College with a degree in biology; her final project focused on a fungal disease afflicting North American bats. As the world's first unvaccinated survivor of rabies, she sees herself as a public figure who can make a difference in the global antirabies effort. On her YouTube channel, she has posted numerous homemade videos to "demonstrate the importance of being rabies-aware," and she maintains a Facebook account—"Jeanna Rabies-survivor Giese."

How could Jeanna Giese have possibly survived? In an article in the *New England Journal of Medicine,* published a year after her release from the hospital, Willoughby (together with seven collaborators) spelled out various unique features of Giese's case that may have aided her survival—such as her youth, her athleticism, and the fact that her exposure to rabies consisted of only a small, superficial puncture on an extremity of her body. They also acknowledged that since viral antigen had never been recovered from her tissues or from the attacking bat, it was possible that Giese was infected with a weak or variant strain of rabies. Nevertheless, the report caused a stir within the small community of rabies experts, who greeted the news of Giese's cure with a mixture of hope and skepticism.

Beyond the rarefied academic laboratories and clinics of the world's top rabies scientists—for whom the apparent benefits of Willoughby's therapy pose an unresolved but largely theoretical question—embattled clinicians, laboring in hospitals big and small, wealthy and poor, are occasionally faced with the question of how to help the patient who arrives already dying of this seemingly indomitable disease. Just as Willoughby discovered during the stressful hours leading up to Giese's diagnosis, treatment options available for rabies are limited to the failed and the unproven. And so, over the past five years, various physicians have begun to attempt Willoughby's controversial method. Now called the Milwaukee protocol, the induction of coma in a rabies sufferer has great appeal when there are no other even anecdotally successful therapies to try.

Unlike in Pasteur's day, when collaboration among doctors usually had to take place under the same roof, today's medical innovators can use technology to share their insights at a distance. On a Web site hosted by the Medical College of Wisconsin, clinicians can download the detailed protocol along with an itemized checklist to guide treatment of a patient with rabies. They also can register the outcome of a case that they treated with the protocol. Willoughby declares that he interprets the data reported here in what he sees as the most conservative way possible, using the principle of "intention to treat." That is, when he tallies the number of times that the Milwaukee protocol has been employed, he includes all of the cases—even those in which the protocol wasn't followed closely (as when a hospital didn't have on hand a drug or monitoring tool essential to the protocol), was interrupted (as when a family removed their relative from care due to concerns about expense), or was applied to patients without normally functioning immune systems (as when immunosuppressed transplant recipients were infected through donated organs).

By this metric, the Milwaukee protocol has been attempted some thirty-five times to date and counts six survivors, including Jeanna. Four of them have not recovered nearly so well as she did: one died of pneumonia before regaining her faculties, and three more are living with profound neurological disabilities. But the most recent (as of this writing, at least) has achieved the best outcome of all. In 2011, Precious Reynolds—an eight-year-old Wiyot Native American from Willow Creek, a small mountain community in far northern California—was diagnosed with an ordinary bout of flu by her local hospital. But soon her grandmother Shirlee Roby got suspicious about her unusual symptoms. "This ain't no damn flu," she exclaimed, and Reynolds was flown more than two hundred miles to UC Davis Children's Hospital.

It turned out that several weeks before her mysterious neurological symptoms appeared, Reynolds (who has been pinning boys to the mat in competitive wrestling ever since the age of four) had tussled with a feral cat outside her elementary school. Based on positive antibody titers of Reynolds's serum and cerebrospinal fluid, rabies was diagnosed, and Willoughby's latest version of the protocol was initiated.

Reynolds remained in a coma for a little over a week, during which time her grandmother stayed by her bedside encouraging her. "I told her she had to put [rabies] on the mat and put him in a half nelson and pin him," Roby said to reporters. "And by golly if she didn't do it." Reynolds left UC Davis Children's Hospital after just fifty-three days of hospitalization, most of them spent in rehabilitation. At her discharge in June 2011, she limped only slightly, supported on her right ankle by a slender brace decorated with butterflies. By summertime, she was playing and swimming with her siblings and cousins; that August, she won twenty-three dollars for her third-place finish in a "mutton bustin'" contest, which involves clinging to the bare back of a sheep as it scampers wildly around a ring.

Back at UC Davis for a checkup in early 2012, Reynolds bounced merrily along the corridors of the pediatric ICU, her butterfly-bedecked brace the only visible remnant of her brush with rabies. She remembers very little of the critical phase of her illness but seems to relish hearing about it from others. She is particularly delighted to hear doctors and nurses confess that they thought she would never survive. As far as Precious is concerned, everything is the same as it was before she got rabies: she still plays soccer, and wrestles, and does generally well at school. She still is fond of animals—most of them, that is. "I don't like cats," she says.

That six out of thirty-five cases have survived after receiving some version of the Milwaukee protocol represents an impressive success rate, at least when compared with the 100 percent fatality rate historically attributed to rabies. Nevertheless, the medical establishment remains largely skeptical. At the time of his original publication, Willoughby declared his intention to set up animal studies to test some of his claims. In particular, the idea that rabies causes mortal complications by way of brain "excitotoxicity"—meaning that the virus overactivates the neurons, disrupting the brain's functions without killing its cells—has yet to be scientifically explored. Six and a half years later, these studies have yet to materialize. The basic reason is financial: Willoughby has not received enough funding to undertake the research himself, and meanwhile no other rabies researcher has made such efforts a priority. Around the world, most of the rabies-treatment research dollars—of which there are very few, given the concentration of the disease in resource-poor areas—are in the hands of scientists who disagree with Willoughby about the underlying biology of rabies infection. Until there is a better understanding of how the rabies virus interacts with the brain on a subcellular level, and of precisely how the various treatments instituted by Willoughby work within the central nervous system, there will be no consensus on the value of the protocol.

Of the Milwaukee protocol's six successful cases, none besides Jeanna's has yet been published in the medical literature. Willoughby has declined to pursue publication of them himself, feeling that the supervising clinicians should be allowed to publish their own observations if they so desire. Meanwhile, in at least two cases where the protocol failed, the physicians have published the results—sometimes with scathing commentary. One such case involved a thirty-three-year-old man treated at the King Chulalongkorn Memorial Hospital in Bangkok, Thailand, home to some of the best-known rabies scientists in the world. The primary clinician on that case, Thiravat Hemachudha, was a vocal skeptic of the protocol before he even tried using it. And in a subsequent paper, he and his colleagues went out of their way to declare that they thought no one should be testing the protocol at all. "There is no credible scientific basis," they write, "for the use of therapeutic coma in rabies, and the risks of this therapy are substantial."

Perhaps the preeminent critic of the Milwaukee protocol is the rabies expert Alan Jackson, who teaches at Queen's University in Kingston, Ontario. Jackson has been a doubter from the very beginning. In the very same issue of the *New England Journal of Medicine* that published the report on Jeanna, Jackson penned a dissenting editorial that cautioned against Willoughby's interpretation of the case. "Induction of coma is not known to have beneficial therapeutic effects in rabies or in other infections of the central nervous system," Jackson averred, adding pointedly: "In the future, induction of coma will probably not be shown to be an effective therapeutic approach to the management of rabies." Even as more apparent successes have emerged, he remains unconvinced, and for an intriguing reason. His central observation about all the survivors, including Jeanna Giese, is that they had significant virus-neutralizing antibodies detectable at the time of diagnosis. This fact points to a robust native immune response, he believes, that might predispose them to survival—regardless of the specific treatments received.

Jackson's argument raises the tantalizing possibility, only hinted at in the margins of the medical literature before Pasteur, that a rare few rabies victims might survive without any intervention at all. We know that this is occasionally true in animals: Pasteur himself recorded the case of a dog that was inoculated with rabies virus, developed neurological symptoms, and then recovered. And since that time, recovery from rabies has been documented in several other animal species, including donkeys, foxes, bats, rats, mice, and guinea pigs.

Giese's survival raises the question of whether this might sometimes happen in humans, too. For more than a hundred years, medical journals have contained occasional case reports that allege survival of rabies. One early nineteenth-century physician claimed in the *Lancet* to have cured rabies by injecting water into a patient's veins; in the middle of the twentieth century, a handful of doctors reported recovery from rabies after they transfused serum from people who were recently vaccinated. A 1972 paper tallied nine cases of reported recovery between 1875 and 1968. A survey for serum rabies antibodies in a population of unvaccinated veterinary personnel yielded some positives, as did a similar study of cave explorers. But because most reports of human rabies survival were made before modern tests for rabies were in routine use, there has been considerable room for doubt as to whether these patients actually suffered from rabies instead of some other malady, real or imagined.

As recently as February 2010, the CDC's *Morbidity and Mortality Weekly Report* published a case report that detailed an apparently unvaccinated survivor of rabies who did not receive the Milwaukee protocol or any intensive care at all. In February 2009, a seventeen-year-old girl appeared at a Texas community hospital complaining of severe headaches, extreme sensitivity to light, vomiting, dizziness, and tingling in the face and arms. When doctors examined her, they found her feverish and disoriented, with a stiff neck typical of nervous-system inflammation. A scan of her brain detected no abnormalities, but a tap of her cerebrospinal fluid showed an increased presence of inflammatory cells. After three days in the hospital, the girl's symptoms had resolved, and she was sent home.

At home, the girl's headaches resumed and intensified. On March 6, she went to another local hospital seeking relief from her headaches and photophobia, which now were occurring alongside muscle aches and pains, particularly in her neck and back. This time her brain imaging and cerebrospinal fluid tap were even more strongly indicative of central nervous system inflammation, and she was transferred to a tertiary-care children's hospital the same day. There, the girl was found to have a variety of signs and symptoms consistent with infectious encephalitis. A flurry of tests ensued to determine a cause. Meanwhile, the girl was treated aggressively for several possible infectious causes of encephalitis: she received antiviral, antibiotic, and antituberculosis drugs.

Despite an extensive workup, no infectious or noninfectious cause for the girl's neurological inflammation was apparent. Then, on March

10, prompted by her doctors, the girl recalled having recently had an encounter with bats. Two months previously, she had entered a cave while on a camping trip and, there, in the dark, had felt the percussive blows of flying bats colliding with her body. She had not noticed any bites or scratches, and she told the doctors that she had never received any rabies vaccine. But the next day, the CDC ran tests and found antibodies against rabies in the girl's blood and cerebrospinal fluid—persuasive evidence of a rabies infection.

On March 14, the girl received a dose of rabies vaccine and immunoglobulin therapy. She remained in the hospital until March 22, receiving only basic supportive care during that time. She made two follow-up visits to the emergency department after her discharge, complaining each time of a recurrence of headaches. On her final visit, she reported relief following a spinal tap. This was the last contact between the girl and her doctors: they record her as being "lost to follow-up." One hopes that medical science will eventually reconnect with this potentially historic case of rabies survivorship.

Is Willoughby correct in his belief that the Milwaukee protocol, refined over time through repeated use, could eventually serve as a reliable treatment—one that transforms rabies from a death sentence into a frequently survivable disease? Or is Alan Jackson correct in his conviction that a very few, fortunate individuals are just naturally predisposed to fight off the disease? Only years of further experimentation can answer these questions definitively. In the meantime, though, we can agree with the assessment of Dr. Rupprecht, back in 2008, when he cautioned both supporters and detractors of the protocol that "we need to focus more on prevention." The protocol provides hope for patients already infected, he observed, but "the odds of coming out without neurological deficits are remote, even with the best care." Three years later, despite the exciting success of Reynolds, this assessment still stands. It's impossible to imagine that developing-world countries would ever have the resources to deploy controlled-coma therapy on more than a tiny fraction of the more than fifty thousand people who currently die of rabies each year. By contrast, it's surprisingly cost-effective for those countries to prevent rabies, through the mass vaccination of dogs.

In our next and final chapter, we consider the ins and outs of canine vaccination but in the dramatic context of a real-world crisis: an outbreak on the Indonesian tourist island of Bali, which until 2008 had been entirely rabies-free.

Vaccination campaign in Jembrana, Bali, 2010.

8

ISLAND OF THE MAD DOGS

Looking back to the beginning—past the scores of Balinese dead from rabies, past the thousands of dogs brutally put down, all the way back to the cold forensic facts of who bit whom—it was probably Thomas Aquino's dog that caused all the trouble. In May 2008, Aquino and a friend, known only as Freddy, sailed to Bali from their native island of Flores, which protrudes from the sea a few hundred miles east down the chain of Indonesian islands that spray off from Singapore toward Australia. Like many travelers on those waters, the two men brought along a dog, the company of which is held to protect the sailor not only from pirates but from more spiritual dangers, too—the mysteries of the animistic Hindu faith extending beyond the iconic beaches and into the deeps of the Bali Sea. For all recorded history the island of Bali had been rabies-free, and so, at least in principle and according to law, dogs could be imported only from other nations where the disease lacked a foothold. But enforcement of this law, fairly reliable at the airport, was entirely absent on Bali's shores. Dogs trotted freely off ferries, pleasure craft, and fishing boats, with no one examining their papers or scanning them for signs of disease. Once on land, these four-footed immigrants mixed easily with the other dogs of the waterfront, where strays (unusual across Bali as a whole, where more than 95 percent of dogs are owned) stalk the beaches in a friendly crush, living off handouts from tourists. The locals, for their part, paid little attention. Among the Balinese it is typical not to acknowledge any dogs except for their own, and sometimes those only cursorily.

So the arrival of Thomas Aquino's dog—silently harboring in its body, somewhere along its nervous system, the seeds of an island-wide epidemic—went wholly unnoticed by anyone, or at least by anyone who understood the scope of the damage that one imported dog could do.

The dog and the two men from Flores settled in Ungasan Village, on the arid Bukit peninsula that hangs like a hammerhead from Bali's southern tip. There, nuclear families make house in compact, densely constructed dwellings, rather than in the sprawling extended-family compounds that are common on the rest of the island. All of Bali thrives on tourism, which every year brings in some two million foreigners and $2.8 billion of their money in search of South Pacific paradise. But lately that business has particularly boomed in Bukit, where large resorts—owned by such luxury conglomerates as InterContinental, Four Seasons, Ritz-Carlton, and Orient-Express—have proliferated. Jobs in and around these resorts have lured many more Indonesians to the peninsula. Muslim and Hindu families live side by side in its neighborhoods, as transplants from the northern part of Bali mix with immigrants from other islands, all seeking to profit from the incessant, explosive growth.

Two months after Thomas's dog made landfall, it bit both him and Freddy. Soon a three-year-old in the village, a boy named Ketut Tangkas, was bitten by his own, suddenly furious dog. By that September, rabies had claimed its first human victim on Bali: a forty-six-year-old woman in the village named Putu Linda. Although a postmortem test for rabies came back positive, public-health officials did not follow up on the implications of that fact for more than two months.

In October, another boy in Ungasan—Muhammad Oktav, also three—was bitten on the face in front of his home by a stray. When his mother took him to the hospital and asked about a rabies vaccine, the doctors refused to provide one. Bali was "rabies-free," they pointed out. They sutured up his wounds and sent him home. Over the next few weeks, the boy seemed to be recovering normally from his injury. He went back to playing as usual in his family's tiled front garden. But a month following the bite, Muhammad abruptly fell ill with chills and, soon, the hallmark fear of water; within two days he was dead. It took two more deaths in Ungasan before the government finally acknowledged, on November 30, 2008, six months after Thomas Aquino and his dog made landfall, that rabies had come to Bali for the first time in history.

One rabid dog had become many. Now, with human lives at stake, no dog could be trusted. With no means of early detection of rabies on the island, the government had recognized the outbreak only after an obvious cluster of human deaths laid bare the extent of the problem. Now the challenge before the Balinese authorities was one of bringing a limited regional epidemic under control before it could spread to the rest of the island.

Barring some miraculous revolution in vaccines, on the one hand, or a near-total obliteration of all animals on the other, the worldwide war against rabies will never be entirely won. Isolated islands like Bali and Britain can win themselves reprieves, at least for a time. But on the larger landmasses, even in nations (such as the United States) where the disease has been largely controlled, its stubborn carriers in wildlife populations reside too far away from man, dispositionally if not always geographically, for us to snuff the devil out very easily. Even if we do somehow succeed in purging rabies from four-legged creatures, we would be left with the problem of the bat, which harbors its own specialized strains of the virus and does not submit easily to the needle.

The great divide in worldwide rabies control is between those nations where the disease has been largely eliminated in *dogs* and those where it has not. This gap is not precisely the same, it should be noted, as the divide between rich and poor. In Brazil, where bat rabies remains rampant but dog rabies has been curtailed through mass vaccination, human deaths from rabies have numbered fewer than ten per year since 2006; in Kazakhstan, which boasts less than a tenth of the population and nearly the same per capita income, the death tally from rabies is significantly higher. The special role of dogs in spreading rabies is due not just to the way they live with us, and all around us; it's also due to the way the virus is perfectly matched to the dog as host, expressing itself in canine saliva at levels rarely achieved in other four-footed wildlife. Rabies coevolved to live in the dog, and the dog coevolved to live with us—and this confluence, the three of us, is far too combustible a thing. According to the CDC, dog bites are still responsible for 90 percent of human exposures to rabies worldwide and more than 99 percent of human deaths.

To be sure, poverty and its attendant ills—governmental corruption, social unrest, poor health overall—help greatly to explain why the two most rabies-afflicted continents, Asia and Africa, have failed to quell the disease. If controlling rabies is in some sense equivalent to keeping dogs healthy, this latter effort is in many ways a proxy for bringing order to civilization itself. Stray dog populations, and rabies along with them, tend to flourish in places where government has broken down; in the former Soviet republics, for example, rabies has resurged during the past twenty years. The problem becomes especially acute under circumstances of radical depopulation. In South Africa's KwaZulu-Natal Province, where some 39 percent of residents are infected with HIV, local veterinarians have reported a tremendous upswing in populations of feral dogs, leading to more cases of rabies. More extreme still, in Ukraine, is the "exclusion zone" around the former Chernobyl power plant, where scientists report an eruption of rabies in the proliferating dogs and other wildlife that haunt the zone.

Even where government functions reasonably well, the peculiarities of rabies often cause officials to become shortsighted. Preventative treatment in dogs often seems like an unaffordable luxury, especially given that the countries where rabies still runs rampant are also countries where other diseases, such as malaria and tuberculosis, kill far more people. But vaccination campaigns in dogs are always much cheaper, over the long haul, than giving out postexposure treatment to humans. The World Health Organization points out that the cost of a full course of postexposure treatment is 4 percent of the average gross national income for Asia and nearly 6 percent of that figure for Africa. As it stands, antirabies campaigns worldwide consume more than a billion dollars per year, and that is just to keep the number of deaths roughly equal to where it stands right now, at approximately fifty-five thousand per year.

As a rabies-free island, Bali was not supposed to add more than 150 victims to that tally. But one wayward dog ensured that it did. Indeed, Bali serves as an instructive story about how quickly any gain in rabies control can be reversed. For all its First World tourist revenue, Bali's per capita income is less than two thousand dollars per year—making residents reluctant to vaccinate at their own expense, and making postexposure treatment hard to afford as well. Just like in the fictional British outbreaks sketched by *The Rage* and *Day of the Mad Dogs,* Bali's rabies-free status wound up becoming a psychological hindrance, as hospitals and residents and governmental officials were too slow to recognize what they were seeing. And when officials did finally awake to the scale of the problem, their response began with the same misguided impulse that tends to strike most government officials when faced with a rabies outbreak: the brute-force killing of dogs.

Dr. Anak Agung Gde Putra is a veterinary epidemiologist at Bali's Disease Investigation Center (DIC), which is not technically part of the government—its funding comes from the United Nations—but which closely advises Bali officials on animal disease control. Dr. Agung is elegant and middle-aged, crisply uniformed in khakis and delicate wire spectacles. His English is excellent, if deliberate, and he talks proudly of the professional visits he has made to Australia and the United States.

On the subject of Bali's rabies outbreak, Dr. Agung felt strongly that his government had acted responsibly given the circumstances. After rabies was confirmed by the laboratory in November 2008, Bali's governor—Made Mangku Pastika, who was elected to his office after becoming well-known as police commissioner for rooting out suspects in the 2002 terrorist nightclub bombing—notified the public within twenty-four hours. The following day he issued a decree laying out the government's course of action, which had been prescribed by Dr. Agung and his colleagues at the DIC. The main thrust of this plan was mass slaughter. It called upon the residents of Bali to kill, personally and by any means necessary, any and all street dogs they came across. In addition, all boats arriving in Bali were to be thoroughly searched by port officials, who would immediately confiscate and destroy any cats, dogs, or monkeys they found.

The governor's early call for citizen-led culling seems never to have caught on, perhaps because the Balinese, overwhelmingly Hindu and great lovers of animals, had little enthusiasm for the job. But government killing of dogs, both unowned and owned, and sometimes without the owners' permission, appears to have proceeded aggressively. Part of the problem was that owners, despite having been told by the government to keep their pets at home during the cull, were accustomed to letting their dogs roam free at all times. Traditional Balinese households own at least one dog, but there is no tradition of confining them; while dogs may be encouraged to remain within the residential compound at night, to protect the family from intruders and evil spirits, they are generally free during the day to forage for food and to socialize among themselves. The typical Bali dog does not wear a collar and has never felt the pull of a leash. So by November 2009, even though only "street dogs" had been officially condemned for depopulation, the government had removed from targeted regions more than the total estimated number of strays in the entire country.

If the numbers were staggering, the methods were ruthless. Pet dogs found at large were shot with pistols or poisoned with strychnine, sometimes whole villages at a time. One particularly shocking bit of footage, which found its way onto YouTube, shows an unidentified man strolling around a Balinese market, dispatching dogs with poisoned blow darts. The dogs stagger for a few paces before they collapse to the ground, writhing and crying as their limbs stiffen. Their faces lock into terrified grimaces as they die. This footage was seized upon and widely circulated by animal-rights groups, not just in Bali but around the world. As one might imagine, this mass extermination campaign began to place a strain on Bali's tourism industry, which still reeled from the bloody Islamist bombings of 2002 and 2005. Six months into the campaign, the *Herald Sun,* Australia's highest-circulating daily newspaper, ran a story about it called "Bali Dog Cull Shocks Aussies." In it, one Australian woman described how her own dog had died after eating a strychnine-laced meatball from a trap. "We found her dead surrounded by vomit and faeces in our garage," the woman said, "and the little meatball was next to her body." Another Australian, a chef, witnessed the shooting of a dog on a beach while a Hindu ceremony took place just nearby.

As the government was eventually to discover, the problem with culling is not that it goes too far but that it can never go far enough. In theory, one could wipe out rabies from a region by exterminating all the dogs. But usually some humans refuse to let that happen, even when a rabies outbreak is charging through their community. Inevitably, there are holdouts: the family with a new puppy, say, or the pensioner whose mutt is not merely his best but his only friend. It takes just a small contingent of softhearted objectors, exempting their own pets, to ruin the whole campaign. Even though many Balinese dog owners seldom come into physical contact with their dogs, often maintaining them in a semi-feral state, as a people they are quite caring and sentimental about their pets. Anecdotally, this is evident as one walks the streets of Ungasan Village, where the outbreak began. Asked about the cull, a young Balinese woman in a T-shirt and plastic flip-flops brags about how she hid her dog and cat in her house. "I love them," she gushed, smiling at a rangy orange tom as it slunk by on the street.

Even in the parts of the world that seem least able to afford a love of animals—places where humans are hungry, where disease runs rampant—this love nevertheless abides. Roughly a third of the world's human rabies cases are believed to occur in India, tempting many officials there to order mass culls. But of course India, too, has its own ancient cultural tradition of preserving animal life. And where rabies is concerned, the more humane alternative is also the more scientifically sound one. In Chennai, India's fifth-largest city, the activist Chinny Krishna of the Blue Cross of India infuriated some officials when he insisted that the local municipality rely on neutering and vaccination to reduce the rabies problem, rather than continuing to cull street dogs. Krishna pointed out that it was in 1860, back when the city was called Madras and ruled by the British, that Chennai first began exterminating dogs in hopes of reducing their number. He says his group became convinced that "if a procedure designed to control or eliminate street dogs had not showed positive results after implementing it for over a hundred years, something was wrong." The rationale of "animal birth control," as Krishna famously called his now-nationwide program—he wanted people to understand that it was "as easy as ABC"—is that neutering and vaccines together will reduce the fraction of dogs susceptible to rabies, creating a stable community of immunized dogs as a barrier to the ongoing spread of the virus.

Bali's initial plan did include some vaccination, in addition to culling. But imported vaccines, which have been proven protective for up to several years with a single dose, were rejected in favor of an inferior vaccine, locally produced in Indonesia, whose average protective effect was less than six months in duration. Moreover, the government chose not to vaccinate island-wide, concentrating its efforts in the area around and to the north of Ungasan Village, with the intention of confining and extinguishing the outbreak on the Bukit peninsula. In both respects, the problem was inadequate funding. According to Dr. Agung at the Disease Investigation Center, when the initial decree responding to the rabies outbreak was released, roughly US$110,000 was allotted for its implementation, but the money wasn't actually made available to those undertaking the task, because of the timing of the end of the fiscal year.

The resulting campaign could never afford to get out in front of the epidemic. Three weeks in, on December 18, the *Jakarta Post* reported that 281 dogs had been destroyed and another 683 vaccinated against rabies. (The government "has vowed to regain Bali's rabies-free status before the end of the year," the newspaper reported.) But by January 9, the government was forced to acknowledge that its efforts to contain rabies on Bukit had failed; a rabid dog had been captured in the capital city, Denpasar. On January 18, scores of high-ranking local government officials participated in a Hindu ceremony at the Puncak Mangu temple seeking divine intervention to stop the outbreak. By November 2009, despite the extermination of 26,705 of Bali's estimated 300,000 dogs and the vaccination of thousands more, the disease had spread to seven of Bali's nine regencies.

Even though postexposure treatment for humans became available in late 2008, people continued to die. Thomas Aquino's friend Freddy wisely began getting shots; Aquino was still deciding whether or not to do so when, on December 14, 2008, he developed muscle cramps and soon began literally foaming at the mouth. Meanwhile, his three-year-old neighbor Ketut Tangkas died at home on December 30.

Although it may have been easier on the government's budget over the short term, and may also have quieted the early Balinese popular outcry for swift, strong action against the outbreak, Bali's decision to slaughter dogs by the thousands rather than concentrate on vaccination—and effective vaccination—proved to be quite expensive in the long run. In its race to win back the island, Bali had given rabies a yearlong head start.

Into this horror story stepped an unlikely demon slayer. In 1973, fresh out of the University of Oregon, Janice Girardi relocated to Bali and began making and selling her own jewelry. By 2007, this operation had swelled to become a multinational business, furnishing pieces to shops and large department stores around the world, and the income allowed Girardi to start a group called Bali Animal Welfare Association (BAWA). BAWA makes its headquarters in the same building in Ubud—the cultural heart of Bali, and a popular destination for the less surf-inclined tourist—that houses the jewelry business. Over the years it has gradually grown to incorporate a fully staffed shelter and veterinary clinic, a twenty-four-hour animal ambulance, a mobile sterilization clinic, a school-based education program, a puppy and kitten adoption program, and a continually expanding range of community programs funded through local and international donations.

The clinic, in particular, stands as a visible monument to Girardi's dedication. Situated in front of lush rice paddies on Ubud's outskirts, it's a graceful two-story building with wooden doors carved in the typical Indonesian style, elaborate hand-cut reliefs of flowers and vines with human and animal figures festooned throughout. The clinic's hallways and terraces are packed full of wire kennels housing softly bedded puppies, which fill the air with their desperate murmurings; ever since rabies came to Bali, the clinic has quarantined all incoming puppies and kittens for one month or more, in order to screen them for signs of rabies. From deeper inside the clinic, the low and earnest barking of more mature dogs adds a subtle baritone to the chorus. BAWA staff move cheerfully about, freshening up cage linens and water bowls and doling out dog food—a mix of rice, carrots, egg, and commercial dog kibble that looks almost appetizing to the human visitor.

In late 2009, after rabies on Bali had started to claim human lives and the mass extermination of dogs had begun, Girardi felt moved to get involved. "At the beginning," she recalls, "I went to meetings where there were hundreds of people clapping when they talked about shooting the dogs or strychnining the dogs. And I'm the only one in the room saying, 'Let's vaccinate!'" A manic talker, Girardi unself-consciously reenacts the scene in her quick staccato, with a chipper grin plastered on her face and her hand stretched high in pantomime of an eager schoolgirl. After some persistence, she persuaded the government to allow BAWA to establish its own vaccination pilot program across the Gianyar regency, which encompasses Ubud and stretches down to greet the island's southeastern shore. Unlike the government, Girardi proposed to use only long-acting foreign vaccines and to kill only those animals that had already demonstrated clear signs of disease. Based on the advice of international rabies experts—from Chulalongkorn University in Bangkok, the World Health Organization in Geneva, and the U.S. Centers for Disease Control and Prevention—she argued that the vaccination would need to cover 70 percent of Gianyar's dogs in order to curb the disease. The campaign proved Girardi correct: such a prevalence of immune dogs, or "warrior dogs," as she later took to calling them, saw the incidence of rabies decline notably in the region.

Despite this success, Girardi had surprising difficulty in convincing the government to extend this campaign island-wide. BAWA played host to a series of international rabies conferences, bringing together Balinese government officials and the world's top rabies scientists; without fail, the latter cited overwhelming evidence in favor of a long-acting vaccine-based strategy for eliminating rabies from Bali, as opposed to large-scale culling. The organization even secured major funding for the larger campaign from the World Society for the Protection of Animals (WSPA), a U.K.-based alliance, and the necessary vaccine from AusAID, the Australian government's foreign-aid program. Still, months of negotiation were required to convince Bali's governor to sign a document approving the plan. In October 2010, more than two years after the first human death from rabies on Bali, the island-wide vaccination effort finally got under way.

The entire campaign was to be coordinated out of BAWA's headquarters in Ubud, a boxy, two-story office structure whose upstairs conference room soon became commandeered as a sort of vaccination war room. There, Girardi and a handful of her BAWA staff—sometimes with Elly Hiby, the London-based head of the Companion Animals Programmes Department for WSPA—could be found in a constant succession of logistical meetings, often huddled over a hand-sketched map of Bali's nine regencies. Into each regency they penciled the numbers of vaccine and surveillance staff required, along with the projected dates. Arrows displayed how the teams would move from regency to regency in pursuit of that 70 percent vaccination rate. This difficult, dangerous work would have to be accomplished neighborhood by neighborhood, compound by compound, dog by dog.

In November 2010, a few weeks into the campaign, the vaccine teams were finishing up the Jembrana regency in rural west Bali, far from the tourist centers of the southern part of the island. Unlike the winding, urban maze of Ubud—where the streets, with their rows of sturdy old family compounds, often feel like fortified lines of walled keeps—Jembrana is more spacious, more agricultural. Some compounds are lined with open metal fencing, so the traveler can catch a glimpse inside; others are hardly fenced at all. The family temples, which in Ubud are graceful concrete structures with elaborately thatched roofs, in Jembrana can sometimes be more ad hoc affairs: piles of bricks, even, with a piece of tin perched on top.

"We are BAWA, here to vaccinate your animals for rabies!" This was the usual exclamation with which the team members entered a family compound. The shouting was necessary in order to be heard above the urgent barking of countless dogs, as the family dogs joined voices with those outside the compound walls: a piercing dissonance of woofs and wails. Made Suwana, BAWA's director of educational outreach, wound up screaming himself as he translated the vaccine team's shouts.

Each vaccine team was made up of four net-wielding dogcatchers; a veterinarian, in charge of drawing up and administering each rabies vaccine; and a record keeper, who noted details about every vaccine recipient on a clipboard. During most of their field excursions, the vaccine team was accompanied by the local *klian banjar,* or elected community leader, who smilingly reassured families of the benign nature of the intrusion. Upon entering each compound, the team asked the residents whether they could handle their own dogs during the injection. The large majority could not. Although the dogs live peacefully among the humans—eating the plentiful remains of the religious offerings laid out daily by the observant women of the family, drowsing comfortably below the *bale bengong* (a sort of gazebo), where the family lounges together during the humid afternoons, or following eagerly at the master's heels as he walks across the road to converse with a neighbor tinkering with his motorbike—the dogs do not approach the family members directly for caresses or for morsels of food, and the family members do not regularly have occasion to lay a hand on the dogs. Indeed, they are usually afraid to do so. As the vaccine team worked, many owners seemed to derive a thrill from watching their semi-wild dogs get unprecedentedly manhandled.

Except in those unusual cases where owners could hold their dogs for the injection, the dogs had to be ensnared in nets. It was a remarkable ballet. As the dogcatchers entered a compound, they fanned out slowly, preparing to corral each dog in turn. A capture was made when one catcher startled a dog toward the other catchers' nets. Except, that is, when the catchers missed: on occasion a net scooped only air, as a wily dog scampered to one side and then sprinted off to some distant corner. Worse, many of the dogs proved capable of breaching the compound fence, escaping into a neighbor's yard or, farther off still, to the impossible catching grounds of a palm forest or rice paddy.

Once a dog was in a net, the net was twirled, such that the dog was left at the net's bottom, twisted into a knot of quivering muscle, fur, and teeth. The dogcatcher then pressed the hoop, with the net now spiraled taut across it, down over the enmeshed dog on the ground. Only then could the veterinarian, working warily through the tangle of ropes, administer the injection of rabies vaccine into the shuddering back muscles of the shrieking animal. Before the dog was released, two measures were taken to identify it as vaccinated. First, by means of long forceps, a red ribbon collar was woven through the net and knotted carefully around the dog's neck. Second, red spray paint was applied generously to the dog's back.

In their demeanor the dogcatchers, generally married men in their early twenties, strived for a nonchalant badassery. They wore their paw print–emblazoned BAWA T-shirts with pants that were either very tight or very loose, along with such rocker accessories as spiked bracelets or bandannas. Most had visible tattoos. During breaks they smoked cigarettes, consumed sweets bought liberally from the ubiquitous household storefronts, and hooted at attractive girls whenever they passed by. When they were engaged in the thrill of the catch, though, the young men's swagger gave way to a quick and purposeful gait; their expressions, coolly bored a moment before, brightened to an alert apprehension.

The most dangerous step, they explained, was the liberation of the dog, which at that moment was often inclined—perhaps understandably—to reel around on its captor, teeth bared. Standard procedure for the release was to hold the hoop at maximum arm's length, with the dog hanging in the net as it gradually untwisted. The dog writhed and snarled in the loose net until a quick flip, judiciously timed, deposited it on the ground. As soon as the dog disentangled its legs, it would be up and sprinting. The only question was, which way? The catcher held up his hoop like a shield until he could be sure the dog was running away from, and not toward, him.

While the catchers demonstrated their derring-do, the record keeper stood beside each owner, earnestly scribbling down the official data—

the name of the owner, the sex of the dog, the dog's name, the dog's age, and the color of the dog's coat—on his clipboard. Not every Balinese dog has a name, but the list of names from one Jembrana community tended toward the punchy and masculine: Kiki, Jos, Boi, Boss, Lupi, Bobo, Inul, Bruno. The sex of more than 80 percent of Balinese dogs is male, due to a common practice of abandoning young female puppies. Some of them survive as strays, but most of them seem to vanish from the island; this practice, though a bit barbaric, has served Bali as a crude form of population control.

A mongrel breed reportedly related to Australia's dingo, the "Bali dog" comes in a variety of colors, from brown to brindle to mottled white. It ranges in size from that of a large beagle to that of a small retriever, with a more or less consistent short stiff coat, erect ears, conical muzzle, and a lean, muscular body. The recent documentary *Bali: Island of the Dogs,* written and hosted by Dr. Lawrence Blair, a garrulous British expat in an eye patch, marshals an impressive group of scholars to testify to the Bali dog's genetic uniqueness. One geneticist at the University of California at Davis, Niels Pedersen, even gives some credence to the legend that one group of wild Balinese dogs, the Kintamani of the interior highlands, is descended from a retinue of chow chows that was imported by an eleventh-century Chinese princess. As the geneticist demonstrates on a "family tree," the Kintamani is very closely related to the chow chow—though he also holds out the possibility that the chow chow might have evolved from the Kintamani, rather than the other way around. Regardless, the Balinese seem convinced that their dogs are noble not merely in temperament but in bloodline.

Among locals, the Bali dog is held to possess nearly sacred properties. In addition to supplying owners with protection—including a sense for metaphysical danger that owners tout as a "magic alarm"—Bali dogs are believed, according to a paper coauthored by Dr. Agung of the DIC, to "cure certain diseases" and more generally to "avert calamity." Sometimes they are given a role in religious ceremonies.

Their apparent ability to survive on rice, the primary foodstuff of the ubiquitous animistic Hindu offering, is frequently cited as evidence of their pluck and fortitude.

Such superstitions are enough to make Putu Ernawati, the smiley young veterinarian on the vaccine team in Jembrana, cautious in predicting the outcome of the island-wide effort. "It is hard to make the village people understand how important the rabies vaccine is," she cautioned. But all around her, there seemed to be a growing awareness of the benefits. On seeing BAWA enter a nearby compound, neighbors would call and wave to make sure they would get visited, too. A few even tried to catch their own dogs in advance of BAWA's arrival—and, having caught them, would advance toward the team with the thrashing, howling dogs in arms, presenting them proudly for injection. One local, Putu Widiasmadi, stood near the front of his compound, clearly enjoying the spectacle of the dogcatchers' exertions. The team record keeper, who was helping to round up dogs, had entrusted his clipboard to two of Widiasmadi's daughters, who laughed uproariously at the names the neighbors had given their pets; apparently, the names of their own dogs, Fred and Ricky, seemed thoroughly reasonable to them by comparison. "I think it's good the government is responding this way to rabies," Widiasmadi said. "Balinese families want to have a dog for protection."

According to one *klian banjar*, the government dog exterminators had come through the village just weeks beforehand. The community still teemed with freewheeling Bali dogs, but soon it became obvious that their owners were steeled to shield them from harm. At one compound, an owner came running at the catchers wielding a large knife and shouting: "No, no, stop! Don't kill my dogs!" At another, a little boy who saw the advancing team ran ahead to the community temple, in order to pray for his dogs' survival.

The dogs themselves kept barking and barking: advancing and barking, retreating and barking; barking as they saw the vaccine team approaching; and barking just as emphatically at the backs of the team as it moved on to the next compound. (One could understand why the Balinese are so supremely confident in the dogs' abilities as protectors; the dogs will bark at anything.) BAWA's teams would provoke a similar din in several more Jembrana communities that same day, and scores more that week. They would need to carry that on week after week, month after month, community after community, regency after regency—until the whole island had barked itself hoarse.

Back at headquarters, on a brilliantly sunny Wednesday, Girardi and her team were strategizing in the war room. In addition to Gianyar, vaccinated during the pilot program, Jembrana was now nearly finished; but the remainder of the island's dogs awaited protection. And while many more teams were currently in training, none were as yet ready to deploy. On a crude hand-drawn map, chopped into rough approximations of the regencies, the team played with numbers. What if they had two teams here, and six teams here? And then, by the next month, ten additional teams?

Deny Gunawan, BAWA's emergency response coordinator, interrupted the meeting to tell Girardi about a call from the clinic. A vicious young dog, which reportedly bit both its owner and its owner's son without provocation, had just been dropped off for examination.

"Was the dog vaccinated?" asked Girardi.

"Not yet," replied Gunawan. He went on to detail two ominous observations that had been made by clinic staff. First, when the dog was caught, it had tried to bite the net. Second, it had run fearfully from the water offered it.

Girardi was unimpressed. These behaviors were typical of a Bali dog when captured and confined. Rather than order the dog's immediate euthanasia, Girardi instructed that it be placed in isolation and monitored for additional symptoms of rabies.

"The dog bit the owner and the son," Gunawan repeated, to be sure that the gravity of the situation had impressed itself on his boss. Girardi, in response, repeated herself as well. The dog was to go into isolation. Girardi wanted to obtain a more complete history from the owner regarding the circumstances of the bites. "Sometimes when you talk to them," she explained to us after Gunawan had gone, "the story will turn out to be, the child was trying to take the toy from the dog and then the dad walked in"—that is, sometimes a biting dog is just being a normal Bali dog, not a mad dog. Besides, she continued, there wasn't anyone in the clinic right now who could properly evaluate the dog.

Girardi returned her attention to the map. Her original plan, which made the most sense from an epidemiological point of view, had been to carry out an organized sweep of the island, starting in Jembrana on the narrow western end and then slowly moving across from west to east, gaining new teams as the island widened. It had become clear, though, that this methodical plan would not be tenable politically. Whenever a new human death or cluster of deaths occurred in some regency, she was immediately pressured to focus her eradication efforts there.

Irrational as those requests were, she was forced to comply. Doing so was necessary not just to keep her good standing with the government; it also was crucial to the animals' welfare. "If we don't get people in there," she pointed out, "they're going to start killing dogs." So an organized eastward campaign across the island had instead given way to something that mapped more like nuclear war: teams would drop in targeted spots and then spread like fallout from the epicenters, until the whole island was consumed.

For all the terror of rabies, for all the superstitions that still attach to the disease (and to the dogs that carry it), and for all the individual intransigence and bureaucratic ineptitude that can mar any response to disease, the theory of vaccination should work: given the right math, and enough vaccine, this most ancient of killers will eventually submit and roll over. But getting the formula right takes patience and political will. BAWA's first pass at vaccinating the island would eventually succeed, essentially as planned. The full array of teams would deploy within a month, and by March 2011 they would hit their target of

70 percent vaccination. The government would eventually fund a second pass, beginning two months later—a crucial step, given Bali's staggering canine turnover rate, with 47 percent of dogs under two years of age. But following several human deaths, apparently from bites that had occurred prior to the BAWA campaign, the government's confidence in vaccination seemed to waver. The culling of healthy dogs resumed, particularly near the recent rabies cases, even though most of the dogs in those communities had already been vaccinated; this once again lowered the overall canine vaccination rate. In May 2011, government officials announced they were abandoning the goal of eradicating rabies on the island by 2012 in favor of the significantly less ambitious deadline of 2015. If Bali recommits itself to building and sustaining its army of "warrior dogs," in Girardi's noble phrasing, then peace should return to this island paradise by then.

After the meeting, Janice drove over to the clinic in her Jeep. Thirty years on the island had assimilated her to the pell-mell driving style of the natives. She zipped around motorbikes, dodged dogs in the road, all at a velocity that was rather unnerving to an American visitor. On arriving, she set out to find the dog that had caused such a stir. It was nighttime, and only a dim fluorescent bulb lit the spot where it was being kept, in a little black wire kennel perhaps two feet wide. For all the fuss about rabies in Bali, you can spend a week reporting on the vaccination campaign—to say nothing of two years researching a book on rabies, or ten years as a U.S. veterinarian—without actually witnessing a rabid dog in the flesh. But here, finally, was one: its head wrenched back on its neck, eyes rolling morbidly in their sockets.

This was not a Bali dog. It was a little "breed dog" (as the locals called them), a black-and-tan Pekingese. It stumbled about like an angry drunk, attacking its cage bars and yowling—a long, mournful, strangled-sounding howl, ending in a wet desperate gurgle. Periodically, the poor dog would slump over in a haze, as if finally spent from its mad exertions. But all of a sudden it would start up again, groaning and snarling, stumbling and biting. It seemed antagonized by our presence, and yet it never seemed fully to recognize we were there.

It's an odd thing to interact with dogs your whole life and yet never see one laid low by this most ancient of canine curses. And in a strange way, it was less terrifying to see in the flesh than to brood upon as a prospect, a threat, a phantom. Just as the needle has become scarier to us than the bite, the reality of the rabid dog cannot quite measure up to the myth. Far more arresting than its rage is its sickness, its absentness.

Many people forget that in the original children's book version of *Old Yeller,* published in 1956, the dog never even develops symptoms of rabies. We are made to understand that the disease is sweeping through the area—afflicting the family's bull, for example, which "reeled and staggered like he couldn't see where he was going. . . . He scrambled to his feet and came on, grunting and staggering and moaning, heading toward the spring." But after Old Yeller gets bitten while tussling with a clearly rabid wolf, the boy shoots the wolf and then, on the hard-spun advice of his pioneer mother, shoots the dog, too, right there and then. It's all over within just a page and a half of junior-reader, large-print type.

To this day it remains shocking that Walt Disney, of all movie moguls, agreed to make *Old Yeller* into a film—and that he didn't find some way to save the dog. The book's author, Fred Gipson, was even hired to write the screenplay. Bill Anderson, a longtime Disney producer who was vice president of studio operations when the film was made, claimed in an interview that a second, more experienced screenwriter, brought in to work with Gipson, had floated the "gutless approach" of saving the dog.

"But Walt could see the drama," Anderson recalled. "Walt knew that—showman that he was—we had to kill the dog to have a story."

At Disney's instigation, though, the writers did make a subtle but extremely important change. To the boy's last-ditch suggestion in the book—that they "tie him or shut him up in the corncrib or some place until we know for sure"—the mother in the film assents, though reluctantly. The dog is confined until, in fact, it becomes mad, snarling at the children, its foaming mouth clearly visible through the wooden gate. For the film, as Anderson put it, they changed the story such that "there was no out but to kill the dog."

For all the generations of children scarred by the movie—Anderson said that his own daughter, many years later, "still doesn't forgive us for killing the dog," and one critic bluntly called the ending "child abuse"—one has to see Walt Disney's subtle amendment as a strange sort of mercy. As brutal as it was for kids to see that mad dog get shot, how much crueler the world would have seemed if its killing had been necessary only because it *might* go mad. Rabies, frightening though its ravages may be to see up close, has about it the comfort of certainty. However much the furious dog was once loved, it cannot be saved.

Dogs' bond with humans is bred into their very cells, their genes; it's written through their entire history, a chronicle that can be read in their eyes. But inside this black wire cage, in the lolling eyes of what remained of a Pekingese, there was nothing legible at all. One could hardly grieve for the dog, because the dog was already gone. To euthanize it—which a BAWA vet mercifully did, moments later, with the customary overdose of anesthesia—was merely to acknowledge its departure.

Hercules Capturing Cerberus by Hans Sebald Beham, 1545.

CONCLUSION

THE DEVIL, LEASHED

Walking the streets of Manhattan, one encounters spitting, staggering, and convulsing on a daily basis, but the offenders generally perambulate on two legs, not four. In July 2009, though, a suspiciously erratic animal was spotted on Payson Avenue, just outside Inwood Hill Park at the island's northern tip. It tested positive for rabies: not an unprecedented occurrence in the borough, but one that officials at the health department felt they should monitor closely. Later that summer, on August 27, 2009, a second case was found, this time along the northern border of Central Park. Then, in December, after more than four months of quiet, ten more were found to be infected, eight of them in Central Park. As 2010 began, the situation started to spiral out of control: twenty-two gone mad in January, twenty-nine in February, with two people bitten. New York's favorite dog-walking destination had quite suddenly become infested with rabies. Yet all the beloved poodles and French bulldogs and bichons frises were not vectors but potential victims.

The perpetrator was a wild animal, but one that nevertheless lives nearly as close among us as our dogs do. Most North Americans, upon being awoken by the characteristic rattle and crash of a trash can lid, will guess the culprit immediately. It is *Procyon lotor,* or the common raccoon, identifiable by its masked face, its shuffling gait, its light- and dark-ringed tuft of tail. During the past twenty years, raccoons have become the U.S. species most frequently found to be infected with rabies. And although their range covers nearly the whole continent, from Canada to Panama, raccoons thrive more today in urban and suburban locales than they do in pristine forests; the highest density of raccoons in New York State is not in the Catskills or the Adirondacks but in New York City. They come, like many human visitors to New York, for the food. One of the most omnivorous animals on earth, the raccoon eats everything—earthworms and acorns, minnows and bluebird nestlings—but it especially enjoys trash, from corncobs to cat chow to cheesecakes. Urban environments, inhospitable to much of the wild kingdom, offer up tons of delectable food for raccoons every day, conveniently presented to them in plastic bags or in cylindrical serving containers that their nimble hands can pry open easily. Sated at this buffet, raccoons can bunk down just about anyplace, whether an abandoned burrow or a sewer, a treetop or a fire escape.

The spread of rabies among raccoons has been described by the CDC as "one of the most intensive wildlife rabies outbreaks in history," and a human blunder seems to have been at least partly to blame. Before the mid-1970s, raccoon rabies—a strain adapted to the raccoon, such that the animal is generally able to survive infection long enough to transmit the virus to another—was confined to Florida and neighboring states, spreading very slowly through the Deep South at a rate of eighteen to twenty-four miles per year. But starting in 1977, more than thirty-five hundred raccoons were legally trapped in Florida and shipped to private hunting clubs in Virginia, where they were released as prospective game. Seeing as how raccoons do not respect the lines of private property, the creatures were effectively being released into the wild. The mid-Atlantic saw its first case of raccoon rabies in 1977, near the Virginia–West Virginia line, far from the existing margins of the southern outbreak. Now there were two loci of infection: one emanating out from Florida, another from the Virginias. Within three decades, rabid raccoons had been observed throughout the entire eastern United States, with a few cases sighted as far west as Ohio.

By 2009, raccoon rabies had already hunkered down in the New York suburbs; it even kept a semipermanent residence in the Bronx. Now, finally, it had swept furiously into Manhattan. Every week, as the city made updates to its online rabies map, the red dots—denoting spots where afflicted raccoons were found—sprayed across Central Park and the vicinity like bullet wounds from a tommy gun. On February 16, 2010, with a total of forty cases identified, the city health department announced a daring plan. It was a scheme that involved, as one rabies expert puts it, "thinking like a raccoon." Based on an understanding of their habitats and behaviors, teams would set collapsible cage traps around the affected areas at night, baiting them with food. The traps would then be checked at dawn and the rabid animals destroyed; those that appeared healthy would be sedated, tagged, vaccinated, and released, on the principle (just as with feral dogs in Bali and elsewhere) that moving or killing healthy animals only invites the population to fill out again with potentially sick ones. Central Park being one of the most popular parks in the world, traps needed to be kept away from the curious hands of children and noses of dogs. Thus the teams had to set traps in out-of-the-way locales, working quickly under cover of darkness: set, sedate, vaccinate, collapse the traps, and go.

Even as this heroic undertaking got under way, rabid raccoons were staggering out with frightening regularity through the green spaces of upper Manhattan, even venturing into the tony neighborhoods nearby. March saw nearly as many cases as February, and April saw nearly as many as March. But by that summer, the reported cases had slowed to a coon-like creep. In the eight months beginning on December 1, 2009, there had been 124 rabid raccoons confirmed, but in July, August, and September 2010 there were only 6; since then, as of the fall of 2011, only 1 more case has been reported. For all the arduousness of trapping and vaccination, the effort seems to have paid off, with the urban jungle of Manhattan left to its more customary ravages.

How, then, to understand the great Manhattan Rabies Outbreak of 2009–10? It was a far cry from the rabies paranoias that racked New York and the great cities of Europe near the end of the nineteenth century. All the way into the twentieth, with Pasteur's cure on the march, rabies lurked as a constant menace in streets and lanes, even in the developed world. The agent of infection was the familiar—the all-too-friendly dog, sometimes even one's own pet—and the consequence of a bite was routinely a horrible death. By contrast, at the beginning of the twenty-first, our dogs are largely vaccinated, our bites are nearly always treated; the threat, meanwhile, lingers among creatures not adapted to our companionship, such as raccoons and foxes, skunks and bats. Rabies has receded to become a sort of spectral presence, a ghost story. It's gone until it isn't.

As we saw in the last two chapters, the struggle of science against rabies seems to have arrived at a frustrating stalemate. Through his innovative, last-ditch protocol for human cases, Rodney Willoughby has broken the 100 percent fatality rate of the world's deadliest virus. But reliable rescue of patients through his methods has proven elusive. By the same token, in the fight to control the exposure rate, the path for government health agencies is clear: with diligent and mandatory vaccination of dogs, human cases can be brought down to near-negligible levels, even in poorer countries where strays are a problem. And yet as the tale of Bali makes clear, beating back the virus can be extremely costly, in both capital and effort, and gains can be quickly reversed if government vigilance flags.

Despite the terrible manner in which it kills, and despite the fact that prevention through dog vaccination is far more cost-effective than treating bites after they happen (to say nothing of managing the fatal cases)—despite all this, the small number of human deaths from rabies means that money for prevention is hard to come by. Charles

Rupprecht, chief of the rabies unit at the CDC and still arguably the world's preeminent expert on the virus, calls this the "cycle of neglected diseases." Rupprecht is an odd combination of a soft speaker and a tough talker, rarely conversing for more than ten minutes without touting his New Jersey roots. Growing up in Trenton, he cultivated a deep love for animals, but also for science fiction, and so he sees the course of his career as having been in some sense inevitable. "When you put the two together—reality and fantasy, animals and monsters—you can't get much closer than rabies," he says with a chuckle. One particularly vivid childhood memory is a face-to-face encounter with a bat, which his brother had inadvertently scared out into view by throwing a basketball against the side of their house. While in graduate school in zoology at the University of Wisconsin, he studied bat ecology; that took him to a summer fellowship in Panama, where he studied bats at the Smithsonian Tropical Research Institute. While he was in veterinary school at the University of Pennsylvania, his interest in bat-borne disease brought him to the Wistar Institute, a biomedical research group on campus, where—his curiosity piqued by the garish medical oddities in the building's front windows—Rupprecht knocked on the door of Tad Wiktor, head of the institute's rabies unit, and introduced himself as a bat specialist. "Ah, bats," Wiktor replied gravely, sucking on his pipe. "I've always wanted bats. How do you *grow* them?"

Since then, Rupprecht has had a hand in nearly every important advance in rabies management during the past twenty-five years. At Wistar, he played a role in developing a recombinant oral rabies vaccine for wildlife, which helped beat back a nasty outbreak among Texas coyotes in the 1990s. He was involved with the first trap-vaccinate-release campaigns among raccoons, on the mid-Atlantic's Delmarva Peninsula and in Philadelphia's Fairmount Park. It was Rupprecht, after moving to the CDC, who consulted with Willoughby when he was scheming a rescue plan for Jeanna; he also was one of the experts that Janice Girardi consulted in devising BAWA's vaccination plan. ("The CDC is a service organization," explains Rupprecht. "It's like *The X-Files*.") It's charming to talk to a virus hunter who speaks with such deep respect for his quarry. An "exquisite parasite," Rupprecht calls rabies, citing all the ways that it has perfectly adapted to its mammalian hosts. "We love to lick, we love to suck, we love to bite," he points out, and so rabies "exploits what mammals do naturally."

Like many other rabies experts, Rupprecht believes that dog rabies—the type responsible for the vast majority of human deaths worldwide—can be eradicated during his lifetime. Unfortunately, though, any concerted effort to do so would require a fairly precise understanding of what the rabies "burden" is: an epidemiologically sound tally of canine infection (which often goes unnoticed or unreported) and human death from rabies (which, if counted at all, often get misclassified as some other encephalitis). This census never gets made, precisely because the number of definitive infections and deaths doesn't seem to justify the spending. "We don't have an effective surveillance system, because there are no resources," Rupprecht says with exasperation. "And there are no resources because there are no cases!" Together with his peers at other health departments worldwide, along with a nonprofit called Global Alliance for Rabies Control, Rupprecht has helped to popularize World Rabies Day, an annual affair designed to build awareness and keep health officials informed about strategies against the disease. If you'd like to mark your calendar, the date each year is September 28—in honor of Louis Pasteur, who died that day in 1895.

There is one realm in which one might say rabies has been conquered during the past ten years—even, in a strange way, enslaved. If this most ancient of viruses can never be eradicated from animals, molecular biologists have hit upon the next best outcome: they are harnessing its uniquely diabolical properties in an attempt to resolve one of our thorniest medical problems. Rabies still knows how to infect us, but at the molecular level we have learned how to infect it.

To see how this is possible, we need to understand what neuroscientists call the "blood-brain barrier." This barrier is not, as the word might imply, a solid wall or even a discrete membrane. Capillaries, just as they do elsewhere in the body, feed every cell in the brain with blood directly; there are nearly four hundred miles of capillaries in the brain alone, each lying less than two-hundredths of an inch from one another. These brain capillaries, however, prohibit most types of molecules from passing through their walls. Oxygen, carbon dioxide, and hormones make it in and out, but larger bodies—including, thankfully, most pathogens—are unable to enter easily.

The existence of the blood-brain barrier was first hinted at by the work of Paul Ehrlich, a German biologist best known for discovering a cure for syphilis. (An unlikely 1940 Hollywood film about that feat, *Dr. Ehrlich's Magic Bullet,* starred Edward G. Robinson as the good doctor.) In dyeing experiments on animals, Ehrlich showed that certain pigments, introduced into the bloodstream, would soon stain all the internal organs—except the brain. Soon, one of his students thought to carry out the opposite side of the equation: injecting the same dye directly into the brain. What he discovered was that the dye stained the whole brain but nothing else. Something unique was happening at the boundary between brain and body.

From the perspective of contemporary medicine, the great irony of the blood-brain barrier is that its parsimony—which, in most cases, protects the brain admirably from infection—becomes a choke hold when the brain falls ill. The barrier does loosen somewhat in situations of infection, but most of the body's own immune responders still have trouble getting in. So, too, do most of the pharmaceutical innovations that have saved untold millions from infectious disease. Antibiotics cannot cross the border, meaning that bacteria such as streptococci, easily snuffed out elsewhere in the body, become fatal meningitis in those uncommon cases when they slip into the brain's membrane. The same is true for antivirals: for example, the herpes virus, a common culprit in viral encephalitis, responds well to antiviral drugs for bodily infections, but these drugs do not pass readily to the brain.

A tremendous focus of pharmaceutical research today is on

finding ways to deliver drug therapies through the barrier. Most of the approaches are in their infancy at this point, so it's hard to say which will prove effective at delivering which therapies, or whether any will wind up being practical to manufacture at the necessary scale to become a regular tool in medicine. One promising vehicle is nanoparticles—meaning particles less than one ten-thousandth of a centimeter in diameter, or less than a quarter the size of the smallest bacteria—which entirely by dint of their diminutive form can slip through the barrier's defenses. Barring that under-the-radar approach, drugs would essentially have to *fool* the barrier, by arriving while attached to some other molecule that naturally passes through the vigilant vessels. That is, they need to hitch a ride with something that itself knows, deep down in its molecular structure, how to slip across the border.

Of all the mechanisms for crossing the blood-brain divide, by far the most surprising—the use of rabies—was dreamed up by a research team led by Priti Kumar, currently an assistant professor at Yale Medical School. Kumar became an innovative biochemist only after a comical series of bureaucratic snafus during her school years. As an undergraduate in Bombay, she intended to study physics but could only get a spot in chemistry. She pursued a concentration in physical chemistry, even finishing two years of a three-year program—only to be told that she hadn't taken enough mathematics in the first year, and so she had to finish her concentration in organic chemistry. When her family moved back to Bangalore (her father worked for the Life Insurance Corporation of India in a "transferable" job, which means they moved him frequently), and Kumar tried to pursue a master's degree there, the local university told her that spots in organic chemistry were full. She would have to shift her focus once again, this time to biochemistry.

It was an involuntary switch, but it turned out to be a happy one. When Kumar went on for her Ph.D. at Bangalore's Indian Institute of Science, she concentrated on the biochemistry of infectious disease.

For her dissertation, she focused not on rabies but on another fascinating zoonosis: Japanese encephalitis, which is carried to humans by mosquitoes from its reservoirs in pigs and birds. The disease infects some thirty thousand to fifty thousand people each year, most of them in a band of oceanic Asia that arcs counterclockwise from Japan through Southeast Asia and comes to rest over the Indian subcontinent. Like rabies, Japanese encephalitis—along with the other, similar viruses in the same class, called flaviviruses, which also include West Nile virus—is a disease that infects the brain, though unlike rabies it travels through the blood rather than the nerves. For ten to fifteen days after exposure, the virus attempts to cross the barrier; when it succeeds, it takes a devastating course, inflaming and often killing regions of the brain responsible for memory and even locomotion. The case fatality rate after infection, while not approaching that of rabies, nevertheless sits at a formidable 30 percent, and many survivors wind up brain-damaged or even paralyzed.

In her thesis, Kumar studied the immune response of humans to Japanese encephalitis, looking for specific T cells that might help some hosts fight off the virus more effectively than others. At her Yale office—a sparsely appointed room that she accesses through a bewildering warren of tunnels and stairs—she reminisces fondly on her training in India. Kumar feels that even though her university didn't have access to the incredible technology that the best American research institutions do, she nevertheless got a world-class education—in part, *because* of that fact. "Here in the United States, when you want to do an experiment, you can buy a kit for it," she explains. "But in India, we had to do everything from scratch, whether it was plating *E. coli* or growing some other bacterium. You start making solutions by weighing out components. You want sodium chloride, you want LB agar? You weigh out everything and autoclave it. So in terms of raw biological knowledge, Ph.D.'s coming out of places like the Indian Institute of Science know an incredible amount."

After graduation, Kumar hoped to find a research team where she could continue working on flaviviruses. She found a group at Harvard that was studying the way that gene therapies, specifically a technique called RNA interference (RNAi), might help to treat flaviviral infections in mice. RNAi uses specially created chunks of RNA that can suppress, or turn off, the harmful effects of certain genes, including those in viruses. It's no exaggeration to say that RNAi, two pioneers of which were awarded the 2006 Nobel Prize in Medicine, is one of the two or three most promising pharmaceutical innovations in a generation. The technique could prove valuable for treating what Kumar calls "undruggable" diseases, particularly in the brain, where the threat of side effects makes most drugs unworkable; RNAi's cardinal virtue is its incredible specificity, because it can (at least in theory) target the harmful effects of disease at the molecular level while leaving the rest of the brain untouched. Kumar and her team readily found an interfering strand of RNA that reduced the fatality rate in mice infected with Japanese encephalitis—when the therapy was delivered by direct injection into the brain.

For human patients, though, the FDA isn't likely to approve brain injections anytime soon. Drugs can't reach the brain through the temples; Kumar says that the only way to deliver these therapies through direct injection would essentially involve brain surgery, and the complication rates of that approach are prohibitively high. And even if those could be brought down to acceptable levels, one imagines that Western patients today would be put off by the prospect of frequent trepanation, which we tend to associate with the medical ideas of the fifteenth century, not the twenty-first. For RNAi to become workable in the brain, then, it needs to find a way in; it needs, that is, to cross the same blood-brain barrier that confounds so many promising brain therapies.

Kumar and her collaborators started with the idea of attaching their treatment to transferrin, a protein that carries iron through the bloodstream. But then they stumbled across a twenty-five-year-old paper that suggested an even more radical idea. Back in 1982, a Yale researcher named Thomas Lentz, in collaboration with three colleagues, showed that rabies took a very particular path into the nervous system: it bound to a specific molecule in peripheral nerves, something called the nicotinic acetylcholine receptor. The receptor is called nicotinic because it serves as the mode by which nicotine makes its way to the brain; it's also the path taken by cobra venom in killing its victim. By linking rabies to this receptor as well, Lentz's work demonstrated for the first time, at a molecular level, the way that rabies so efficiently worked itself into the nerves. More than that, though, in a subsequent paper eight years later, Lentz went so far as to isolate *which part* of the rabies virus accomplished this trick: a particular peptide, made of twenty-nine amino acids, that bound the virus to the receptor.

If this rabies peptide used the receptor in the peripheral nerves, Kumar and her team reasoned, it might be able to exploit the same receptor at the boundary to the brain. They started with Lentz's peptide and then refined it. They attached it to fluorescents in order to show that it could penetrate the brain; sections of mouse brain showed that the peptide did, in fact, carry the dye into the entire brain. Finally, they assembled a treatment molecule to deliver the RNA therapy, with this crucial section of the rabies virus—a key, as it were, to unlocking the door to the brain—out in front. What they found was impressive: after treatment with the molecule, 80 percent of the mice fought off the infection of Japanese encephalitis, compared with none of the control group. And this success has been replicated: three years later, in March 2011, a team at Oxford further refined their carrier molecule and thereby delivered large quantities of an anti-Alzheimer's RNAi to the brains of mouse subjects.

It's far too early, of course, to declare victory against the blood-brain barrier, or to declare rabies the agent of its conquest. After all, countless thousands of mice are "saved" every year by drugs that will never see the inside of a person, let alone preserve a human life. There is not yet even a single FDA-approved drug that employs RNAi technology; the closest to market is probably a drug to fight macular edema (that is, swelling) in diabetics, which lingered in Phase II trials as of March 2011.

But Kumar's triumph in the laboratory, besides giving hope for treatment of brain illnesses in general, presents two grand, historical ironies—not noteworthy, perhaps, in the context of contemporary science, but germane to the four-thousand-year acquaintance of humans with rabies. The first is that rabies, for so long our most visible and intractable animal-to-human infection, could be harnessed in the treatment of another deadly zoonosis, namely Japanese encephalitis. After we have spent millennia weathering maladies derived from pigs and fowl, it is sweet revenge to think that we might use rabies to combat some of these diseases in the twenty-first century.

The second and even more gratifying irony is the method by which rabies has been exploited: the hollowing out of the virus for the use of its shell—the possession of it, one might say. As we have seen, rabies itself is our most ancient possessor, devouring the brains of its victims, transforming them into slavering vehicles for its own malign spread. Its evolutionary strategy, maximally fatal, must also be maximally manipulative: given only a brief window to replicate itself, the virus must incite its hosts to stalk and to salivate, to obsess and to attack. That humans die, and die terribly—with the otherworldly aversion to water, the hallucinations, the foaming and gulping, and worse—is just a senseless side effect.

If we can never completely eradicate rabies, we can at least take some comfort in the fact that we now have turned this cruel gambit back on itself. Just as rabies exploited the sociability of dogs in aiding its spread, humanity has now taken rabies' own defining characteristic—its efficient binding into the nervous system—and seized control over it, in the hopes of saving human lives. We have charmed the beast, mesmerized it, forced it to do our bidding. One is reminded of Orpheus, who, in search of his dead love Eurydice, employed his beautiful music to retrieve her from the underworld. "Cerberus stood agape," records the poet, "and his triple jaws forgot to bark."

ACKNOWLEDGMENTS

First, we must acknowledge the great debt we owe to the historians and scientists who have written extensively during the past fifty years on different aspects of rabies: Jean Théodoridès, Patrice Debré, Neil Pemberton and Michael Worboys, Kathleen Kete, Harriet Ritvo, Alan C. Jackson, John D. Blaisdell, Wu Yuhong, Merritt Clifton, Bert Hansen, George M. Baer, Charles Rupprecht, and others.

We are also enormously indebted to Neil Henry and the Berkeley Graduate School of Journalism for their generous appointment of us as visiting scholars, which allowed us to use the world-class resources of the University of California at Berkeley to complete our research on this book. Many thanks, too, to the UC libraries and their staff.

Thanks are due, as well, to friends and colleagues who helped us with research and inspiration during the writing of this book. Jon Lackman, Gideon Lewis-Kraus, and Henrik Kuhlmann gave indispensable translation advice. Ellen Silbergeld's lab at the Johns Hopkins School of Public Health provided an early sounding board for the book's basic structure. *Wired* gave Bill leave to finish this project; thanks in particular to Thomas Goetz, Jake Young, and Chris Anderson. Also, many thanks to Rafil Kroll-Zaidi, Meghan Davis, and Jess

Benko, all of whom read early drafts of the book and gave valuable feedback.

On a personal note, we'd like to thank our family: Bob and Mary Wasik, Emmett L. Murphy, Jean Austin, Dave and Jen Wasik, Becca Hurley, Emmett J. and Anne Murphy, and Janet Wasik.

We're tremendously grateful to our agent, Tina Bennett, for encouraging us in this project and helping to see it through. Thanks, too, to the brilliant team at Viking who made this book crisp and beautiful: Maggie Riggs, Wendy Wolf, Bruce Giffords, Ingrid Sterner, Jim Tierney, Carolyn Coleburn, Rebecca Lang, Yen Cheong, Kevin Doughten, and all the rest.

Finally, we can't say enough about our editor and friend Josh Kendall, who championed this book and improved it at every step of the process. From his deep interest in Greek myth to his obsession with cheap horror, Josh's enthusiasm for this project has been nothing short of rabid.

NOTES

INTRODUCTION: LOOKING THE DEVIL IN THE EYE

Page

1 **Consider the kamikaze bobcat in Cottonwood:** John Faherty, "No Tall Tale: Rabid Bobcat Invades Cottonwood Bar," *Arizona Republic,* March 27, 2009.

1 **Or the frenzied otter in Vero Beach:** Associated Press, May 27, 2007.

1 **Or the enraged beaver in the Loch Raven Reservoir:** Bob Allen, *North County News,* Aug. 22, 2007.

2 **one young couple in the Adirondack hamlet:** Bob Condon, "Man Bitten in Attack by Rabid Fox," *Glens Falls Post-Star,* April 9, 2008.

2 **"What disturbs me," remarked one Connecticut man:** NBC Connecticut, Aug. 18, 2010.

2 **For one victim in Putnam County:** *This American Life,* Oct. 27, 2006.

2 **the red fox in South Carolina:** Jill Coley, "Deadly, Grisly Rabies Still Threat in Lowcountry," *Charleston Post and Courier,* Nov. 20, 2008.

3 **a dog was attacked by a mad peccary:** KPHO.com, Feb. 18, 2011.

3 **a skunk that beset the pet Pekingese**: ThePilot.com, Nov. 24, 2010.

3 **a donkey fell prey to the madness:** Brennan Leathers, "Donkey Bite Prompts Rabies Warning," *Post-Searchlight,* Jan. 12, 2011.

3 **In Imperial, Nebraska, the afflicted animal:** Russ Pankonin, "Rabid Sheep Causes Stir at the Fair," *Imperial Republican,* Sept. 3, 2010.

4 **"the lethal gift of livestock":** Jared Diamond, *Guns, Germs, and Steel* (New York: W. W. Norton, 2003), 195.

4 **"breathed out nastier germs":** Ibid., 195.

5 **Susan Sontag noted that even as late as:** *Illness as Metaphor and AIDS and Its Metaphors* (New York: Picador, 2001), 126–27. *Illness as Metaphor* was originally published in 1978; *AIDS and Its Metaphors,* from which this particular insight was drawn, was originally published in 1988.

6 ***King of the Hill:*** "To Kill a Ladybird," December 12, 1999.

6 ***Beavis and Butt-Head:*** "Rabies Scare," March 18, 1994.

6 ***Scrubs:*** "My Lunch," April 25, 2006.

6 ***The Office:*** "Fun Run," September 27, 2007.

6 **fifty-five thousand, in the estimate:** World Health Organization, September 2011.

7 **a portrait of Lennox as a boy:** George Romney, *Charles Lennox, 4th Duke of Richmond, Duke of Lennox, and of Aubigny* (ca. 1776–77).

7 **so named in honor of Gebhard Leberecht von Blücher:** Rosemary Baird, *Goodwood: Art and Architecture, Sport and Family* (London: Frances Lincoln, 2007), 170.

8 **it began one day with shoulder pains:** Alan C. Jackson, "The Fatal Neurologic Illness of the Fourth Duke of Richmond in Canada: Rabies," *Annals of the RCPSC* 27, no. 1 (Feb. 1994).

8 **On YouTube one can find video:** http://www.youtube.com/watch?v=OtiytblJzQc.

8 **"I don't know how it is":** Frederic Tolfrey, *The Sportsman in Canada, Vol. 2* (London: T. C. Newby, 1845), 228.

9 **The next day, the duke ate and drank:** Jackson, "Fatal Neurologic Illness."

9 **so repelled was he by the water in the basin:** Baird, *Goodwood,* 170.

9 **"the patient is seized with sudden terror":** Armand Trousseau, *Lectures on Clinical Medicine* (Philadelphia: Lindsay & Blakiston, 1873), 2:85.

9 **sometimes occurring at a rate of once per hour:** *Morbidity and Mortality Weekly Report* 59, no. 38 (2010): 1236–38.

9 **an unfortunate porter who suffered such emissions:** Armand Trousseau, *Lectures on Clinical Medicine* (Philadelphia: Lindsay & Blakiston, 1867), 1:686.

10 **the duke dictated a lengthy letter:** Baird, *Goodwood,* 171.

10 **once saw Pasteur perform this trick:** Axel Munthe, *The Story of San Michele* (New York: Dutton, 1930), 51.

10 **"At the beginning of each session":** Patrice Debré, *Louis Pasteur,* trans. Elborg Forster (Baltimore: Johns Hopkins University Press, 1998), 430.

CHAPTER 1: IN THE BEGINNING

15 **"Mix us stronger drink":** *The Iliad of Homer,* trans. Richmond Lattimore (Chicago: University of Chicago Press, 1997), 9.203.

15 **"I have learned to be valiant":** Ibid., 6.444.

16 **"a hunting hound in the speed of his feet":** Ibid., 8.338–39.

16 **his pitch to Achilles:** Ibid., 9.238–39.

16 **no fewer than nine terms:** Thomas Walsh, *Fighting Words and Feuding Words: Anger and the Homeric Poems* (Lanham, Md.: Lexington Books, 2005), 1–3.

16 **It has not been invoked anywhere in the poem:** The definitive accounting of *lyssa* in Homer is Bruce Lincoln's essay "Homeric *Lyssa:* 'Wolfish Rage,'" in *Death, War, and Sacrifice: Studies in Ideology and Practice* (Chicago: University of Chicago Press, 1991), 131–37.

16 **goading Heracles to slay his family:** Euripides, *Heracles.*

16 **Pentheus's own mother and aunt:** Euripides, *The Bacchae.*

16 **a feminine form wearing a dog's head:** One such image is here: http://www.theoi.com/Gallery/N17.1.html.

17 **"as a snake waits for a man by his hole":** *Iliad of Homer,* trans. Lattimore, 22.93.

17 **"powerful *lyssa* unrelentingly possesses":** *Iliad* 21:242–43; translated in Lincoln, "Homeric *Lyssa:* 'Wolfish Rage.'"

17 **one of humanity's first recorded jokes:** The joke (as well as its interpretation) comes to us from Andrew R. George, "Ninurta-pāqidāt's Dog Bite, and Notes on Other Comic Tales," *Iraq* 55 (1993): 63–75.

19 **"If a dog becomes rabid":** Wu Yuhong, "Rabies and Rabid Dogs in Sumerian and Akkadian Literature," *Journal of the American Oriental Society* 121, no. 1 (Jan.–March 2001): 33.

19 **"Like a rabid dog, he does not know":** Ibid.

19 **the omens of entrails readers:** Ibid., 35.

19 **lunar eclipses in particular months:** Ibid., 36.

19 **the Marduk Prophecy:** Ibid., 37–38.

19 **some of the incantations:** Ibid., 38–42.

21 **The *Samhita* devotes nearly a thousand words:** Kaviraj Kunja Lal Bhishagratna, trans., *An English Translation of the Sushruta Samhita* (Calcutta: published by the author, 1907), 733–36.

21 **"The bodily Vāyu, in conjunction":** Ibid., 733–34.

23 **his notes on hydrophobia:** Celsus, *De medicina,* book 5, chap. 27.

23 **an anonymous methodist text:** Ivan Garofalo, ed., *De Morbis Acutis et Chroniis* (New York: E. J. Brill, 1997), 85–89.

23 **the notes on hydrophobia made by Soranus:** *On Acute Diseases and On Chronic Diseases,* trans. I. E. Drabkin (Chicago: University of Chicago Press, 1950), 361–89.

24 **"The victims of hydrophobia die quickly":** Ibid., 367.

24 **Based on findings of teeth and bones:** Karen Rhea Nemet-Nejat, *Daily Life in Ancient Mesopotamia* (Westport, Conn.: Greenwood Press, 1998), 111.

25 **dating back as far as 3500 B.C.:** Katharine Rogers, *First Friend: A History of Dogs and Humans* (New York: St. Martin's Press, 2005), 29.

25 **when archaeologists excavated her temple at Isin:** Jeremy Black and Anthony Green, *Gods, Demons, and Symbols of Ancient Mesopotamia* (Austin: University of Texas Press, 1992), 70.

25 **often scarred with knife marks:** Nicholas Wade, "In Taming Dogs, Humans May Have Sought a Meal," *New York Times,* Sept. 8, 2009.

26 **"come to the world of men in the shape":** Willem Bollée, *Gone to the Dogs in Ancient India* (Munich: Verlag der Bayerischen Akademie der Wissenschaften, 2006), 54.

26 **rife with images of dogs as battlefield scavengers:** Ibid., 33–34.

26 **An excavated tomb at Abydos:** Michael Rice, *Swifter Than the Arrow: The Golden Hunting Hounds of Ancient Egypt* (New York: I. B. Tauris, 2006), 36–37, 46, 55.

27 **References to dogs as scavengers in Egypt:** D. M. Dixon, "A Note on Some Scavengers of Ancient Egypt," *World Archaeology* 21, no. 2 (Oct. 1989): 193–97.

27 **"They were afraid that some *lyssa*":** Xenophon, *Anabasis,* book 5, chap. 7.

28 **"My dogs in front of my doorway":** *The Iliad of Homer,* trans. Richmond Lattimore (Chicago: University of Chicago Press, 1997), 22.66–76.

29 **Iris slinging it at Athena:** R. H. A. Merlen, *De Canibus: Dog and Hound in Antiquity* (London: J. A. Allen, 1971), 27.

29 **renders both transitions with awful acuity:** Ovid, *Metamorphoses,* trans. Charles Martin (New York: W. W. Norton, 2010), book 3, 252–318.

30 **As described by Hesiod, Cerberus was quite friendly:** Hesiod, *Theogony,* trans. Richmond Lattimore (Ann Arbor: University of Michigan Press, 1959), 770–74.

30 **"slaver from Cerberus":** Ovid, *Metamorphoses,* book 4, 683.

30 **along with a creation myth:** Ibid., book 7, 578–95.

30 **symptoms of aconite poisoning:** John Blaisdell, "A Frightful, but Not Necessarily Fatal, Madness: Rabies in Eighteenth-Century England and English North America" (Ph.D. diss., Iowa State University, 1995), 18.

31 **In 2001, two researchers at France's Institut Pasteur:** Hassan Badrane and Noël Tordo, "Host Switching in *Lyssavirus* History fro[illegible] Chiroptera to the Carnivora Orders," *Journal of Virology* 75, no. 17 (2001): 8090–104.

32 **a team led by the Stanford epidemiologist:** Nathan Wolfe, "The Origin of Malaria: Discovered," *Huffington Post,* Aug. 3, 2009.

32 **particularly intriguing details about smallpox:** Yu Li et al., "On the Origin of Smallpox: Correlating Variola Phylogenics with Historical Smallpox Records," *PNAS* 104, no. 40 (2007): 15787–92.

32 **archaeological evidence shows the clear presence:** Donald Hopkins, *The Greatest Killer: Smallpox in History* (Chicago: University of Chicago Press, 2002), 14–15.

34 **the disease does appear in Ge Hong's:** Joseph Needham, *Science and Civilisation in China* (Cambridge: Cambridge University Press, 2000), VI:6:91–92.

34 **Things totter off the rails with Pliny:** Pliny the Elder, *Natural History,* book 29, chap. 32.

CHAPTER 2: THE MIDDLE RAGES

40 **as the historian John Cummins notes:** John Cummins, *The Hound and the Hawk: The Art of Medieval Hunting* (New York: St. Martin's Press, 1988), 74.

40 **stag pursued by hounds would sometimes figure:** Brigitte Resl, ed., *A Cultural History of Animals in the Medieval Age* (New York: Berg, 2007), 76.

40 **one Christian allegorist likened the ten points:** Cummins, *Hound and the Hawk,* 68.

40 **Bestiaries, in their treatment of the stag:** Resl, *Cultural History of Animals in the Medieval Age,* 61.

40 **One fourteenth-century German work:** Cummins, *Hound and the Hawk,* 72.

41 **ordered not to eat the flesh of wild beasts:** Exodus 22:31.

42 **"his hidden parts were made rotten and stinking":** Matthew Zimmern, "Hagiography and the Cult of Saints in the Diocese of Liège, c. 700–980" (Ph.D. diss., University of St. Andrews, 2007), 48.

43 **Hubert's own set of otherworldly interventions:** Ibid., 51.

43 **petitioned the current bishop, Waltcaud:** Satoshi Tada, "The Creation of a Religious Centre: Christianisation in the Diocese of Liège in the Carolingian Period," *Journal of Ecclesiastical History* 54, no. 2 (April 2003): 218–19.

46 **"The *chien baut* must not give up":** Cummins, *Hound and the Hawk,* 19.

46 **The Castilian king Alfonso XI:** Ibid., 25.

46 **One medieval archbishop of Canterbury:** George Jesse, *Researches into the History of the British Dog* (London: Robert Hardwicke, 1866), 2:36.

47 **Thomas à Becket:** Ibid., 2:38.

47 **"A grehounde sholde be heeded lyke":** Ibid., 2:136–37.

47 **William of Wykeham upbraided one particular abbey:** Eileen Power, *Medieval People* (London: Methuen, 1950), 121.

48 **list of English public records:** Jesse, *History of the British Dog,* 2:7.

48 **In other accounts, peasants who took game:** Matt Cartmill, *A View to a Death in the Morning: Hunting and Nature Through History* (Cambridge, Mass.: Harvard University Press, 1993), 61.

48 **standard practice for all commoners' dogs:** Jesse, *Researches into the History of the British Dog,* 1:375.

48 **"many dead throughout the city":** Philip Ziegler, *The Black Death* (New York: Penguin, 1982), 58.

48 **one chronicle reports grave diggers:** Joseph Patrick Byrne, *The Black Death* (Westport, Conn.: Greenwood Press, 2004), 108.

49 **"as if they were mounds of hay":** Joseph Patrick Byrne, *Daily Life During the Black Death* (Westport, Conn.: Greenwood Press, 2006), 101.

49 **In painted plague scenes:** Christine Boeckl, *Images of Plague and Pestilence: Iconology and Iconography* (Kirksville, Mo.: Truman State University Press, 2000), 64.

49 **the expression "six feet under":** Ibid., 16.

50 **The preeminent Arab physician:** Anna Campbell, *The Black Death and Men of Learning* (New York: AMS Press, 1966), 35.

50 **a tract theorizing that the new:** Ibid., 47–51.
50 **Gentile of Foligno thought:** Ibid., 53–55.
50 **A tractate from the medical faculty:** Ibid., 55–58.
50 **Alfonso of Córdoba likewise blamed:** Ibid., 52–53.
51 **a physician from the French town:** Ibid., 60–62.
51 **"festering boils . . . break out on people":** Exodus 9:8–9.
52 **a massive swine flu epidemic:** Francisco Guerra, "The Earliest American Epidemic: The Influenza of 1493," *Social Science History* 12, no. 3 (Fall 1988): 313–19.
52 **put the death toll from disease in Hispaniola:** Mary Ellen Snodgrass, *World Epidemics* (Jefferson, N.C.: McFarland, 2003), 51.
52 **"the cause of the ailments so common among us":** Ibid., 50.
52 **an English-language translation and expansion:** Edward, second Duke of York, *The Master of Game* (New York: Duffield, 1909), 85–104.
54 **historians have found annual outlays:** Cummins, *Hound and the Hawk,* 30.
55 **By 1288, a French wag:** From *XMLittré,* an online version (hosted at http://francois.gannaz.free.fr/Littre/) of Émile Littré's 1863 historical dictionary of the French language. Many thanks to Jon Lackman for the interpretation.
55 **by 1678:** Ibid.
56 **"is that illness is *not* a metaphor":** Susan Sontag, *Illness as Metaphor and AIDS and Its Metaphors* (New York: Picador, 2001), 3.
56 **Through assiduous translation to Arabic:** Peter E. Pormann and Emilie Savage-Smith, *Medieval Islamic Medicine* (Cairo: American University Press in Cairo, 2007), 16, 83, 97–98.
57 **By the tenth century, Baghdad:** Ibid., 83.
57 **the thirteenth century saw the establishment:** Ibid., 97–98.
57 **a process akin to scholarly peer review:** Ray Spier, "The History of the Peer-Review Process," *Trends in Biotechnology* 20, no. 8 (Aug. 2002): 357.
57 **"There was with us in hospital":** Pormann and Savage-Smith, *Medieval Islamic Medicine,* 116.
57 **His preferred treatment for bites:** Jean Théodoridès, *Histoire de la rage: Cave canem* (Paris: Masson, 1986), 48. Thanks to Alex Bedrosyan of the Columbia University Tutoring and Translation Agency for the translation.
57 **he anchors his observation with a personal narrative:** Ibid., 50.
58 **the great doctor expressed the belief that heat and cold:** Ibid., 48–49.

58 **A fairly lengthy treatment of rabies:** *The Medical Writings of Moses Maimonides* (Philadelphia: Lippincott, 1963), 1:67–72.

59 **Artifact collectors have preserved:** Pormann and Savage-Smith, *Medieval Islamic Medicine,* 152.

59 **They were blood relatives of Saint Catherine:** Fabián Alejandro Campagne, "Charismatic Healers on Iberian Soil: An Autopsy of a Mythical Complex of Early Modern Spain," *Folklore* 118 (April 2007): 44–46.

59 **In 1619, a shoemaker named Gabriel Monteche:** María Tausiet, "Healing Virtue: Saludadores Versus Witches in Early Modern Spain," *Medical History Supplement* 29 (2009): 47–48.

61 **On two occasions during the 1570s:** William Christian Jr., *Local Religion in Sixteenth-Century Spain* (Princeton, N.J.: Princeton University Press, 1981), 6, 11, 29, 40.

61 ***saludadores* also had a reputation:** Tausiet, "Healing Virtue," 50–51.

CHAPTER 3: A VIRUS WITH TEETH?

65 **Over four densely cited pages:** Juan Gómez-Alonso, "Rabies: A Possible Explanation for the Vampire Legend," *Neurology* 51, no. 3 (1998): 856–59.

65 **Even *Playboy* weighed in:** "Humping Like Rabids," *Playboy,* March 1, 1999.

66 **does raise many intriguing parallels:** Gómez-Alonso, "Rabies."

67 **"nothing was spoken of but vampires":** Voltaire, *A Philosophical Dictionary* (London: John and Henry L. Hunt, 1824), 6:306.

67 **the self-described age of reason:** Wayne Bartlett and Flavia Idriceanu, *Legends of Blood: The Vampire in History and Myth* (Westport, Conn.: Praeger, 2006), 19.

69 **"Frightened, [Lycaon] runs off":** Ovid, *Metamorphoses,* trans. Charles Martin (New York: W. W. Norton, 2010), book 1, 323–32.

69 **Old Norse gives us the legend:** Sabine Baring-Gould, *The Book of Were-Wolves* (New York: Causeway Books, 1973), 39–40.

70 **the Laighne Faelaidh, a race of men:** George Henderson, *Survivals in Belief Among the Celts* (Glasgow: J. Maclehose & Sons, 1911), 170.

70 **A number of ancient Indo-European tribal names:** Ian Woodward, *The Werewolf Delusion* (London: Paddington Press, 1979), 30.

70 **When Herodotus writes of the Neurians:** Baring-Gould, *Book of Were-Wolves,* 9.

70 **an account of a half-human tribe in India:** David Gordon-White, *Myths of the Dog-Man* (Chicago: University of Chicago Press, 1991), 49.

70 **Strabo, a geographer from the first century:** Patricia Dale-Green, *Lore of the Dog* (Boston: Houghton Mifflin, 1967), 170.

70 **Similarly, the Ch'i-tan:** Gordon-White, *Myths of the Dog-Man,* 133.

70 ***cynocephali*, or "dog-headed men":** Ibid., 63.

71 **a taxonomy for the thousands:** Barbara Allen Woods, *The Devil in Dog Form: A Partial Type-Index of Devil Legends* (Berkeley: University of California Press, 1959).

71 **"If there is any merit":** Ibid., 33.

72 **Nicholas Remy turned this same reasoning:** Nicholas Remy, *Demonolatry* (London: J. Rodker, 1930), 70.

72 **moments of particular wickedness:** Woods, *Devil in Dog Form.*

72 **"with strange pleading eyes":** Ibid., 77.

73 **"That which is once forsworn":** Ibid., 113.

73 **Elizabeth Clarke, who during the seventeenth:** Dale-Green, *Lore of the Dog,* 79.

73 **Alison's account of her dog's attack:** E. Lynn Linton, *Witch Stories* (London: Chapman and Hall, 1861), 270.

74 ***1521.* Two admitted werewolves:** Bartlett and Idriceanu, *Legends of Blood,* 94.

74 ***1530.* Near Poitiers, three enormous wolves:** Montague Summers, *The Werewolf* (London: K. Paul, Trench, Trubner, 1933), 225.

74 ***1541.* A farmer in Pavia:** Baring-Gould, *Book of Were-Wolves,* 64–65.

74 ***1558.* Near Apchon:** Summers, *Werewolf,* 228.

75 ***1573.* The town of Dole:** Baring-Gould, *Book of Were-Wolves,* 74–78.

75 ***1598.* An entire family:** Ibid., 78–81.

75 **That same year, near Angers:** Ibid., 81–84.

75–76 ***1603.* Jean Grenier . . . snatched from a cradle:** Ibid., 85–99.

77 **"confessed to me also":** Pierre de Lancre, *On the Inconstancy of Witches,* trans. Gerhild Scholz Williams (Tempe: Arizona Center for Medieval and Renaissance Studies, 2006), 331.

77 **an account of rabies in 1702:** Richard Mead, *A Mechanical Account of Poisons* (London: J. Brindley, 1745), 150–51.

78 **Mead even goes so far:** Ibid., 154–55.

78 **Like many physicians of his day:** Anna Marie Roos, "Luminaries in Medicine: Richard Mead, James Gibbs, and Solar and Lunar Effects on

the Human Body in Early Modern England," *Bulletin of the History of Medicine* 74, no. 3 (Fall 2000).

78 **"varied both in colour and magnitude":** Ibid., 445.

78 **"depended upon the lunar force":** Richard Mead, *A Treatise Concerning the Influence of the Sun and Moon upon Human Bodies* (London: J. Brindley, 1748), 64.

78 **the "legend of the torn garment":** Woods, *Devil in Dog Form,* 95.

79 **a vampire account from Baghdad:** Baring-Gould, *Book of Were-Wolves,* 253.

79 **the remedy for dog bite that Richard Mead:** Mead, *Mechanical Account of Poisons,* 164–65.

80 **"recovered without the help":** Ibid., 178.

80 **"sucking the blood of people and cattle":** Bartlett and Idriceanu, *Legends of Blood,* 12.

80 **the great wave arrived:** Ibid., 13.

81 **"After it had been reported":** From Paul Barber's translation of "Visum et repertum," included in *Vampires, Burial, and Death: Folklore and Reality* (New Haven, Conn.: Yale University Press, 2010), 16.

82 **the release of pent-up gases:** Ibid., 161.

82 **a tale from Siret, in northern Romania:** Matthew Beresford, *From Demons to Dracula: The Creation of the Modern Vampire Myth* (London: Reaktion, 2008), 64.

82 **another folklorist lists the animal forms:** Barber, *Vampires, Burial, and Death,* 87.

83–84 **The proprietor of a hotel across the lake:** Dorothy Hoobler and Thomas Hoobler, *The Monsters: Mary Shelley and the Curse of Frankenstein* (New York: Little, Brown, 2006).

85 **Goaded by a lover, Polidori:** David Lorne Macdonald, *Poor Polidori: A Critical Biography of the Author of* The Vampyre (Toronto: University of Toronto Press, 1991), 95–97.

86 **"Usually they bite at night":** Gonzalo Fernández de Oviedo y Valdés, trans. Sterling Stoudemire, *Natural History of the West Indies* (Chapel Hill: University of North Carolina Press, 1959), 62.

86 **Translations of Oviedo's abridged history:** Kathleen Myers, *Fernández de Oviedo's Chronicle of America: A New History for a New World* (Austin: University of Texas Press, 2007), 3–4.

87 **a 1796 account of his years in Suriname:** J. G. Stedman, *Narrative of a Five Years' Expedition Against the Revolted Negroes of Surinam…* (London: J. Johnson, 1806), 146–47.

88 **Goya was using spectral, bat-like figures:** James Twitchell, *The Living Dead: A Study of the Vampire in Romantic Literature* (Durham, N.C.: Duke University Press, 1981), 20–29.

88 **"whole circumstance has lately been doubted":** Charles Darwin, *The Voyage of the* Beagle, entry for April 9.

CHAPTER 4: CANICIDE

91 **"One cannot conceive," Campbell wrote:** *Millennial Harbinger* 5, no. 1 (1848): 267–69.

92 **Rumor had it that to end:** *Notes and Queries* 6, no. 148 (1852): 207.

92 **as one admirer noted years after her death:** Charles Waterton, *Essays on Natural History: Third Series* (London: Longman, Brown, Green, Longmans, and Roberts, 1857), 177.

92 **"no nose was so much talked of":** Quoted in "Charles Kirkpatrick Sharpe," *Temple Bar: A London Magazine for Town and Country Readers,* July 1889.

93 **An 1830 paper in the *Lancet*:** *Lancet,* Feb. 6, 1830, 619.

93 **"Not only a most disgusting":** Alfred Swaine Taylor, *On Poisons in Relation to Medical Jurisprudence and Medicine* (Philadelphia: Lea & Blanchard, 1848), 457.

93 **another called it "degrading":** *Medical Adviser, and Guide to Health and Long Life,* Oct. 2, 1824, 242.

93 **believed to be some 100,000 pet dogs:** Kathleen Kete, *The Beast in the Boudoir: Petkeeping in Nineteenth-Century Paris* (Berkeley: University of California Press, 1994), 53–54.

94 **"the degraded state and savage disposition":** Charles Darwin, *The Variation of Animals and Plants Under Domestication* (London: John Murray, 1868), 2:46. Harriet Ritvo's splendid book *The Animal Estate* draws out this theme in far more detail.

94 **ten times more likely to die:** Harriet Ritvo, *The Animal Estate: The English and Other Creatures in the Victorian Age* (Cambridge, Mass.: Harvard University Press, 1987), 169–70.

95 **a list of twenty-one supposed causes:** Benjamin Rush, *Medical Inquiries and Observations* (Philadelphia: J. Conrad, 1805), 2:303–5.

95 **On those instances when they did diverge:** Lester King, *The Medical World of the Eighteenth Century* (Chicago: University of Chicago Press, 1959), 59–60.

95 **saw the body mechanistically:** Ibid., 65–83.

96 **From November to May, Rush:** David Hawke Freeman, *Benjamin Rush: Revolutionary Gadfly* (Indianapolis: Bobbs-Merrill, 1971), 48.

96 **"Having dined on beef, peas, and bread":** Ibid., 57–58.

96 **"mischievous effects on the nervous system":** Ibid., 110.

96 **how to make saltpeter:** Ibid., 127, 136.

96 **Rush inoculated Patrick Henry:** Ibid., 130.

97 **on hand to perform battlefield medicine:** Ibid., 178.

97 **a tradition, beginning at least with Boerhaave:** John Blaisdell, "A Frightful, but Not Necessarily Fatal, Madness: Rabies in Eighteenth-Century England and English North America" (Ph.D. diss., Iowa State University, 1995), 36, 39.

97 **"was uncommonly sizy in a boy":** Rush, *Medical Inquiries and Observations,* 311.

97 **his 1792 doctoral thesis on the disease:** James Mease, *An Inaugural Dissertation on the Disease Produced by the Bite of a Mad Dog, or Other Rabid Animal* (Philadelphia: Thomas Dobson, 1792).

98 **a second pamphlet on rabies:** James Mease, *Observations on the Arguments of Professor Rush, in Favour of the Inflammatory Nature of the Disease Produced by the Bite of a Mad Dog* (Whitehall, Eng.: William Young, 1801).

98 **"One of the first things I can remember":** Unpublished autobiographical notes, 2, James Mease Archive, UCLA Biomedical Library.

99 **a comical ditty, called "The Two Dog Shows":** Neil Pemberton and Michael Worboys, *Mad Dogs and Englishmen: Rabies in Britain, 1830–2000* (New York: Palgrave Macmillan, 2007), 70.

99–100 ***"Le chien est une machine à aimer":*** Gordon Stables, *Notre ami le chien* (Paris: J. Rothschild, 1897), 1.

100 **"Great Dog Massacres":** Kathleen Kete, "*La Rage* and the Bourgeoisie: The Cultural Context of Rabies in the French Nineteenth Century," *Representations,* no. 22 (Spring 1988), 90 and n12.

100 **In England, the preferred method of dispatch:** Ritvo, *Animal Estate,* 191–92.

100 **"turned ordinary people into murderers":** Pemberton and Worboys, *Mad Dogs and Englishmen,* 74.

100 **"Constantinople and Africa are rabies-free":** Kete, "*La Rage* and the Bourgeoisie," 97.

100 **reports were coming in from India:** Ritvo, *Animal Estate,* 174.

100 **"exhausted" their "nervous system":** Ibid., 180.

100 **"Hydrophobia makes its appearance":** Pemberton and Worboys, *Mad Dogs and Englishmen,* 31.

100 **Different theories fingered different breeds:** Ritvo, *Animal Estate,* 181.

100 **One letter writer to the London *Times*:** Pemberton and Worboys, *Mad Dogs and Englishmen,* 31.

101 **In the 1850s, France created:** Kete, "*La Rage* and the Bourgeoisie," 100.

101 **Britain had a similar tax:** Ritvo, *Animal Estate,* 179.

101 **"its instinct impels it, at times":** George Fleming, *Rabies and Hydrophobia* (London: Chapman and Hall, 1872), 194.

101 **"invariably express an exaggerated attachment":** Kete, "*La Rage* and the Bourgeoisie," 101.

103 **the strange dog, clearly in distress:** Elizabeth Gaskell, *The Life of Charlotte Brontë* (London: Smith, Elder, 1857), 308–10.

103 **"I doubt whether . . . no harm will ensue":** Charlotte Brontë, *Shirley* (New York: Harper & Brothers, 1850), 451.

104 **"would have been, had she been placed in health":** Gaskell, *Life of Charlotte Brontë,* 302.

104n **"The surprise is not that the Brontës died so young":** Beth Torgerson, *Reading the Brontë Body* (New York: Palgrave Macmillan, 2005), 2–3.

105 **what he calls "biological horror":** Jason Colavito, *Knowing Fear* (Jefferson, N.C.: McFarland, 2008), 78.

105 **"a bizarre liminal creature poised somewhere":** Ibid., 65.

105 **an 1830 letter to the London *Times*:** *Times* (London), June 4, 1830.

106 **"with ape-like fury":** Robert Louis Stevenson, *The Strange Case of Dr. Jekyll and Mr. Hyde* (London: Longmans, Green, and Co., 1886), 37.

106 **"Leaving to the patient all the faculties":** Kete, *Beast in the Boudoir,* 101.

107 **"vacant converse with spectral and imaginary objects":** Letter reprinted in *The Complete Works of Edgar Allan Poe* (New York: The University Society, 1902), 335.

107 **"a violent delirium, resisting the efforts":** Ibid., 336.

108 **R. Michael Benitez developed a theory:** R. Michael Benitez, "Rabies," *Maryland Medical Journal* 45 (1996): 765–69.

109 **a thoroughly dubious 1830 address:** H. W. Dewhurst, *Observations on the Probable Causes of Rabies, or Madness, in the Dog* (London: published for the author, 1831), 9–14.

109 **one generally respected text from 1857:** Kete, *Beast in the Boudoir,* 103.

109 **an 1845 proposal, penned by a certain Monsignor:** "Project for the Prevention of Hydrophobia in Man," translated in *Monthly Journal of Medical Science,* Nov. 1845, 878–79.

110 **In 1830, when the British Parliament:** Pemberton and Worboys, *Mad Dogs and Englishmen,* 24–25.

110 **a tally of rabies experts surveyed by M. J. Bourrel:** Kete, *Beast in the Boudoir,* 105.

111 **"that rabid man related by Haller":** Rossi, trans. Dell'Orto, "Mylabris Fulgurita—Its Use in Hydrophobia," *New Orleans Medical and Surgical Journal* 11 (Jan. 1884): 539.

111 **Bachelet and Froussart emphasize the weakness:** Kete, *Beast in the Boudoir,* 102.

111–12 ***l'enfant du diable,* or "the child of the devil":** William Baillie-Grohman, *Camps in the Rockies* (New York: Charles Scribner's Sons, 1882), 401.

112 **"there is no wild beast in the West":** Theodore Roosevelt, *Hunting Trips of a Ranchman* (New York: G. P. Putnam's Sons, 1884), 33–34.

112 **In the 1870s, when the army colonel:** Richard Irving Dodge, *The Plains of the Great West and Their Inhabitants* (New York: G. P. Putnam's Sons, 1877), 95.

112 **Roosevelt, in one of his memoirs, recalled:** Roosevelt, *Hunting Trips of a Ranchman,* 33–34.

112 **Perhaps the most spectacular attack:** Fred Gowans, *Rocky Mountain Rendezvous: A History of the Fur Trade Rendezvous, 1825–1840* (Salt Lake City: G. M. Smith/Peregrine Smith Books, 1985), 80–95.

113 **Another rabies-addled wolf rampaged:** Dodge, *Plains of the Great West and Their Inhabitants,* 97–98.

114 **"was saved on account of wolf biting through pants":** Benteen to Theodore Goldin, Feb. 22, 1896, in "The Benteen-Goldin Letters," mimeographed (ca. 1952), Bancroft Library, University of California at Berkeley.

114 **quantities of the deadly poison strychnine:** Baillie-Grohman, *Camps in the Rockies,* 406.

114 **The anthropologist George Bird Grinnell:** George Bird Grinnell, *Blackfoot Lodge Tales* (New York: Scribner, 1892), 283.

114 **Colonel Dodge put forward the truly odd claim:** Richard Irving Dodge, *Our Wild Indians* (Hartford, Conn.: A. D. Worthington, 1882), 320–21.

114–15 **one particularly tantalizing Native American cure:** *American Farmer,* Feb. 1, 1828, 367.

115 **"From the colonists' perspective, Indians":** Jon Coleman, *Vicious: Wolves and Men in America* (New Haven, Conn.: Yale University Press, 2004), 42–43.

115 **"wild beasts and beast-like men":** Richard Hildreth, *The History of the United States of America* (New York: Harper & Brothers, 1849), 1:281.

115 **"act like wolves and are to be dealt withal as wolves":** Coleman, *Vicious,* 43.

115 **One tribe, the Skidi Pawnee:** Ibid., 45–46.

116 **"There was not the slightest danger from them":** Francis Parkman, *The Oregon Trail* (Boston: Little, Brown, 1872), 324.

116 **forgoing the purchase of a carriage:** Thomas Brock, *Robert Koch: A Life in Medicine and Bacteriology* (Washington, D.C.: American Society for Microbiology, 1999), 22–23.

116 **began his studies of anthrax in 1873:** Ibid., 31–35.

CHAPTER 5: KING LOUIS

For the essential facts of Pasteur's life we have relied primarily on two sources. The first is the extensive biography penned soon after Pasteur's death by his son-in-law, René Vallery-Radot, and translated into English in 1916 by Mrs. R. L. Devonshire. The second is Patrice Debré's 1995 biography, translated in 2005 by Elborg Forster. Below we cite only specific quotes from these works, as well as facts drawn from other works; uncited facts may be assumed to derive from Vallery-Radot, Debré, or both.

121 **Roux's medical training had been temporarily disrupted:** Hubert Arthur Lechevalier and Morris Solotorovsky, *Three Centuries of Microbiology* (New York: Dover, 1975), 143.

121 **"This Roux is really a pain":** Patrice Debré, *Louis Pasteur,* trans. Elborg Forster (Baltimore: Johns Hopkins University Press, 2005), 334.

122 **"Live in the serene peace":** René Vallery-Radot, *The Life of Pasteur,* trans. Mrs. R. L. Devonshire (Garden City, N.Y.: Doubleday, Page, 1916), 451.

122 **"When I see a child":** Ibid., 447.

122 **"I am now wholly wrapped up":** Ibid., 172.

123 **leading to a case fatality rate:** http://www.nlm.nih.gov/exhibition/smallpox/sp_variolation.html.

123 **within a year the Prince of Wales's daughters:** Abbas M. Behbehani, "The Smallpox Story: Life and Death of an Old Disease," *Microbiological Reviews* 47, no. 4 (1983): 455–509.

123 **Only after the unexpected death of Louis XV:** Frank Fenner et al., *Smallpox and Its Eradication* (Geneva: World Health Organization, 1988), 255.

124 **more than 100,000 were vaccinated:** Sheryl Persson, *Smallpox, Syphilis, and Salvation: Medical Breakthroughs That Changed the World* (Wollombi, NSW, Australia: Exisle, 2009), 31.

124 **scientists and laypeople who claimed:** Arthur Allen, *Vaccine: The Controversial Story of Medicine's Greatest Lifesaver* (New York: W. W. Norton, 2007), 56–57, 64–69.

125 **rampant in France during the 1870s:** Bernard J. Freedman, "A Tale of Two Holidays: How to Make Great Discoveries," *British Medical Journal,* July 15, 1989, 162.

125n **the French historian Antonio Cadeddu:** Gerald L. Geison, *The Private Science of Louis Pasteur* (Princeton, N.J.: Princeton University Press, 1996), 40.

127n **"The Koch group, which relied on":** Thomas D. Brock, *Robert Koch: A Life in Medicine and Bacteriology* (Madison, Wisc.: Science Tech, 1988), 171–72.

128 **"[t]he twenty-five unvaccinated sheep will perish":** Vallery-Radot, *Life of Pasteur,* 315–20.

128 **"As M. Pasteur foretold":** Nigel Kelly, Bob Rees, and Paul Shute, *Medicine Through Time* (Oxford: Heinemann Educational, 2002), 87, quoting *Times* (London), June 3, 1881.

129 **"This malady is one of those":** Debré, *Louis Pasteur,* 417, quoting Emile Roux, "L'oeuvre medicale de Pasteur," *Agenda du Chimiste* (1896).

131 **"absolutely ignorant of any connection":** Vallery-Radot, *Life of Pasteur,* 391.

131 **"This is indeed a new disease":** Debré, *Louis Pasteur,* 419, quoting Pasteur in Vallery-Radot, *Maladies virulentes, virus-vaccins, et prophylaxie de la rage,* in *Oeuvres de Pasteur* (Paris: Masson et Cie, 1939), 555.

132 **Pasteur referred to the unseen:** Debré, *Louis Pasteur,* 414–15.

133 **"'We absolutely have to inoculate the rabbits":** Debré, *Louis Pasteur,* 429, quoting R. Rosset, *Pasteur et la rage* (Lyon: Fondation Mérieux, 1985).

134 **"The seat of the rabic virus":** Vallery-Radot, *Life of Pasteur,* 172.

134 **"It is torture for the experimenter":** Debré, *Louis Pasteur,* 421.

134 **"[Roux], [Charles] Chamberland, and [Louis] Thuillier":** Debré, *Louis Pasteur,* 430, quoting Rosset, *Pasteur et la rage.*

136 **"Until now I have not dared":** Vallery-Radot, *Life of Pasteur,* 404.

137 **"I have not yet dared to treat":** Ibid., 410.

138 **"On 6 July at eight o'clock":** Debré, *Louis Pasteur,* 439.

139 **"My dear children":** Vallery-Radot, *Life of Pasteur,* 416.

139 **"Cured from his wounds":** Ibid., 417.

140 **"Very good news last night":** Ibid.

140 **"Hydrophobia, that dread disease":** Ibid., 422.

141 **"he had a kind word for every one":** Ibid., 447.

141 **"I have such confidence in the preventive forces":** Pasteur Institute Web site, http://www.pasteurfoundation.org/historic.shtml.

142 **"Is this all we have come":** Vallery-Radot, *Life of Pasteur,* 426.

142 **the story was raptly followed:** Bert Hansen, "Medical Advances in Nineteenth-Century America," *History Now,* no. 10 (Dec. 2006), http://www.gilderlehrman.org/historynow/12_2006/historian6.php.

142 **the four of them were trotted out:** Bert Hansen, "America's First Medical Breakthrough: How Popular Excitement About a French Rabies Cure in 1885 Raised New Expectations for Medical Progress," *American Historical Review* 103, no. 2 (1998): 373–418.

142 **Many newspapers also went out:** Hansen, "Medical Advances in Nineteenth-Century America."

143 **"It reversed the assumption":** Ibid.

143 **Some, in the decade or so:** Hansen, "America's First Medical Breakthrough."

143 **"From the heights of our settled situations":** Bruno Latour, *The Pasteurization of France,* trans. Alan Sheridan and John Law (Cambridge, Mass.: Harvard University Press, 1988), 130, quoting Jeanne, "La bactériologie et la profession médicale," *Concours Médicale* 4, no. 5 (1895): 205.

144 **By the year 1900:** Pasteur Institute Web site, http://www.pasteur foundation.org/historic.shtml.

144 **the institute's purposes were:** Debré, *Louis Pasteur,* 467.

145–46 **"Dr. von Frisch . . . has not succeeded":** Ibid., 460.

146 **"How difficult it is to obtain":** Vallery-Radot, *Life of Pasteur,* 433.

146 **"Pasteur continues to be fairly well":** Debré, *Louis Pasteur,* 494.

146 **"Our only consolation, as we feel":** Vallery-Radot, *Life of Pasteur,* 439.

CHAPTER 6: THE ZOONOTIC CENTURY

152 **it was proved beyond doubt that this pathogen:** Dave Mosher, "Black Death's Daddy Was the Bubonic Plague," *Wired Science,* Oct. 8, 2010.

152 **But throughout 1918, during the height of the epidemic:** Alfred W. Crosby, *America's Forgotten Pandemic: The Influenza of 1918* (Cambridge: Cambridge University Press, 2003), 296–97.

152 **reported its toll in stark terms:** J. S. Koen, "A Practical Method for Field Diagnosis of Swine Diseases," *American Journal of Veterinary Medicine* 14 (1919): 469–70.

153 **"I believe I have as much to support this diagnosis":** Ibid., 470.

153 **"a peroration . . . worthy of Luther":** Crosby, *America's Forgotten Pandemic,* 297–98.

154 **some pathbreaking research on distemper:** George Dunkin and Patrick Laidlaw, "Dog Distemper in the Ferret," *Journal of Comparative Pathology and Therapeutics* 39 (1926): 201–12.

154 **In 1933 they succeeded, isolating a virus:** Gina Kolata, *Flu: The Story of the Great Influenza Pandemic of 1918 and the Search for the Virus That Caused It* (New York: Farrar, Straus and Giroux, 1999), 75.

154 **That same year Shope:** Ibid., 76.

154 **"the virus of swine influenza is really the virus":** Patrick Laidlaw, "Epidemic Influenza: A Virus Disease," *Lancet,* May 11, 1935, 1118–24.

155 **a report from late in that decade:** George A. Denison and J. D. Dowling, "Rabies in Birmingham, Alabama," *JAMA,* July 29, 1939, 390–95.

155 **more than 250 deaths were logged:** Ibid.

157 **as the Hurston scholar Robert Haas points out:** Robert Haas, "Might Zora Neale Hurston's Janie Woods Be Dying of Rabies? Considerations from Historical Medicine," *Literature and Medicine* 19, no. 2 (Fall 2000): 209, 211–18.

157 **Hurston's brother and first husband:** Ibid., 209–11.

157 **a more intriguing and ultimately more plausible:** Robert Haas, "*The Story of Louis Pasteur* and the Making of Zora Neale Hurston's *Their Eyes Were Watching God:* A Famous Film Influencing a Famous Novel?" *Literature/Film Quarterly* 32, no. 1 (2004): 12–19.

158 **"Somehow, I talked my mother into taking me":** Stanley Wiater, Matthew R. Bradley, and Paul Stuve, eds., *The Twilight and Other Zones: The Dark Worlds of Richard Matheson* (New York: Citadel Press/Kensington, 2009), 12.

158 **"Those were very bad years":** Douglas Winter, *Faces of Fear: Encounters with the Creators of Modern Horror* (New York: Berkley Books, 1985), 28.

160 **has been appropriated by American fiction:** See W. B. Seabrook, *The Magic Island* (New York: Harcourt, Brace, 1929).

160 **"Anubis":** Paul Gagne, *The Zombies That Ate Pittsburgh* (New York: Dodd, Mead, 1987), 24.

160 **"basically ripped off" Matheson's vision:** Joe Kane, *Night of the Living Dead* (New York: Kensington, 2010), 22.

160 **pooled six hundred dollars apiece:** Gagne, *Zombies That Ate Pittsburgh,* 21, 29–32.

161 **"the post-millennial ghoul of the moment":** Warren St. John, "Market for Zombies? It's Undead (Aaahhh!)" *New York Times,* March 26, 2006.

161 **The sci-fi blog io9.com made a chart:** http://io9.com/5070243/.

161 **zombie booms correlated with Republican rule:** Peter Rowe, "With Obama Election Comes the Return of the Vampire," *San Diego Union-Tribune,* November 8, 2008.

162–63 **The film's director, Danny Boyle, says:** Matthew Hays, "Return of the Killer Zombies!" *Mirror* (Montreal), June 26, 2003.

162n **"Continue the termination. Don't stop believing":** Chuck Klosterman, "How Modern Life Is Like a Zombie Onslaught," *New York Times,* December 3, 2010.

164 **Late one summer morning in 1953:** Homer D. Venters et al., "Rabies in Bats in Florida," *American Journal of Public Health* 44, no. 2 (1954): 182–85.

164 **he remembered something he had read:** *St. Petersburg Times,* July 31, 1960.

164 **Beginning in 1906, ranches in southern Brazil:** Paul W. Clough, "Rabies in Bats," *Annals of Internal Medicine* 42, no. 6 (1955): 1330–34.

165 **In 1911, a São Paulo laboratory:** Aurelio Málaga-Alba, "Vampire Bat as a Carrier of Rabies," *American Journal of Public Health* 44, no. 7 (1954): 909–18.

165 **the scientific community in Brazil was convinced:** David Brown, *Vampiro: The Vampire Bat in Fact and Fantasy* (Silver City, N.M.: High-Lonesome Books, 1994), 72.

165 ***tumbi baba* in Paraguay, *rabia paresiante* in Argentina:** Málaga-Alba, "Vampire Bat as a Carrier of Rabies."

165 **devastation brought by aerial assault:** Victor Carneiro, "Transmission of Rabies by Bats in Latin America," *Bulletin of the World Health Organization* 10 (1954): 775–80.

165 **The first human deaths attributed to rabies:** Holman E. Williams, "Bat Transmitted Paralytic Rabies in Trinidad," *Canadian Veterinary Journal* 1, no. 1 (1960): 20–24.

165 **Since dog rabies had been eliminated:** Carneiro, "Transmission of Rabies by Bats in Latin America."

165 **In the three decades that followed, eighty-nine humans:** Williams, "Bat Transmitted Paralytic Rabies in Trinidad."

165 **In 1951, a Mexican man, prior to succumbing:** Málaga-Alba, "Vampire Bat as a Carrier of Rabies."

166 **By the end of 1965, infected bats:** George M. Baer and Devil Bill Adams, "Rabies in Insectivorous Bats in the United States, 1953–65," *Public Health Reports* 85, no. 7 (1970): 637–46.

166 **today, only Hawaii's bats are rabies-free:** Catherine Brown et al., "Compendium of Animal Rabies Prevention and Control," *Journal of the American Veterinary Medical Association* 239, no. 5 (2011): 609–18.

166 **Bat bites are now the cause:** Web site of the U.S. Centers for Disease Control and Prevention.

166 **anyone who awakens with a bat:** Ibid.

166 **a letter to the *Lancet* in 1983:** Jane Teas, "Could AIDS Agent Be a New Variant of African Swine Fever Virus?" (letter), *Lancet* 321, no. 8330 (1983): 923.

167 **In late 1984, a research team at Harvard:** Mirko Grmek, *History of AIDS: Emergence and Origin of a Modern Pandemic* (Princeton, N.J.: Princeton University Press, 1990), 80–81.

167 **A short June 1987 letter to the *Lancet*:** F. Noireau, "HIV Transmission from Monkey to Man" (letter), *Lancet* 329, no. 8548 (1987), 1498–99.

167 **The following month Abraham Karpas:** Abraham Karpas, "Origin of the AIDS Virus Explained?" *New Scientist,* July 16, 1987.

168 **One AIDS researcher interviewed teens:** Diane Goldstein, *Once Upon a Virus: AIDS Legends and Vernacular Risk Perception* (Logan: Utah State University Press, 2004), 85.

168 **"The first one I heard was about a sailor":** Ibid., 86.

168 **in Scotland, a focus group convened:** Jenny Kitzinger and David Miller, "'African AIDS': The Media and Audience Beliefs," in *AIDS: Rights, Risk, and Reason,* ed. Peter Aggleton, Peter Davies, and Graham Hart (London: Falmer Press, 1992), 40–41.

169 **In 1990, an AIDS reearcher in Punta Gorda:** Stephanie Kane, *AIDS Alibis: Sex, Drugs, and Crime in the Americas* (Philadelphia: Temple University Press, 1998), 55.

169 **In their version, which has circulated:** Heike Behrend, "The Rise of Occult Powers, AIDS, and the Roman Catholic Church in Western Uganda," in *AIDS and Religious Practice in Africa,* ed. Felicitas Becker and P. Wenzel Geissler (Boston: Brill, 2009), 36n9.

169 **traced this myth back to a 1991 story:** *Sunday Mail* (Harare), Sept. 29, 1991, quoted in Alexander Rödlach, *Witches, Westerners, and HIV: AIDS and Cultures of Blame in Africa* (Walnut Creek, Calif.: Left Coast Press, 2006), 160–61.

170 **In some African countries, the white man:** Behrend, "Rise of Occult Powers," 36n9.

171 **Just before the tunnel opened, one poll:** Julian Barnes, *Letters from London* (New York: Vintage, 1995), 288.

171 **in an earlier survey, carried out:** *New York Times,* Dec. 26, 1985.

171 **"The Channel Tunnel is a violation":** Eve Darian-Smith, *Bridging Divides: The Channel Tunnel and English Legal Identity in the New Europe* (Berkeley: University of California Press, 1999), 149.

171 **"the blessing of insularity," one member:** Ibid., 147.

172 **"The commercial began sedately":** S. J. Taylor, *Shock! Horror! The Tabloids in Action* (London: Corgi, 1992), 34–35.

174 **security fences with animal-proof mesh:** Darian-Smith, *Bridging Divides,* 146–48.

174 **its PR handlers revealed to the media:** *New York Times,* Feb. 17, 1994.

174 **"as if lining up behind Mitterrand":** Barnes, *Letters from London,* 287–88.

175 **most recent rabid animal to be unwittingly imported:** BBC News, April 26, 2008.

175 **more than ninety Americans contracted in 2003:** Donald G. McNeil Jr., "Monkeypox Cases Surge in Rural Areas as Price of the Victory over Smallpox," *New York Times,* Aug. 30, 2010.

175 **a survey of 122 human cases in Bangladesh:** Stephen Luby et al., "Recurrent Zoonotic Transmission of Nipah Virus into Humans, Bangladesh, 2001–2007," *Emerging Infectious Diseases* 15, no. 8 (2009).

176 **In Afghanistan, the nation's lone pig:** Reuters, April 30, 2009.

176 **Tunisia went so far as to ban:** News24, Oct. 6, 2009.

176 **among newspaper cartoonists in Muslim countries:** Anti-Defamation League, "Arab Cartoonists Use Swine Flu Theme to Mock Israeli Leaders," *Jewish State,* May 22, 2009.

176 **an Egyptian cleric, Sheikh Ali Osman:** "Fatwa in Egypt: Source of Pigs Is Jews," *Al Bawaba,* May 11, 2009.

177 **Video footage shows workers:** http://www.youtube.com/watch?v=LYB4sDKh3FI.

177 **Other amateur footage showed pigs brained:** Michael Slackman, "Cleaning Cairo, but Taking a Livelihood," *New York Times,* May 25, 2009.

CHAPTER 7: THE SURVIVORS

182 **As she sat beside her mother:** Jeanna Giese's personal Web site, http://site.jeannagiese.com/My_Story.html.

182 **Later, Giese showed the tiny wound:** "The Girl Who Survived Rabies," *Extraordinary People,* Discovery Channel (2006).

187 **it had been shown in a 1992 study:** Brian Paul Lockhart, Noel Tordo, and Henri Tsiang, "Inhibition of Rabies Virus Transcription in Rat Cortical Neurons with the Dissociative Anesthetic Ketamine," *Antimicrobial Agents and Chemotherapy* 36, no. 2 (1992): 1750–55.

188 **At 10:00 p.m. on the evening of October 10, 1970:** Michael A. Hattwick et al., "Recovery from Rabies: A Case Report," *Annals of Internal Medicine* 76, no. 6 (1972): 931–42.

188 **Over the next few days, Winkler's condition:** Ibid.

189 **Although no virus was isolated:** Ibid.

189 **After days spent motionless in a coma:** Ibid.

189 **Winkler's clinicians—led by Dr. Michael A. Hattwick:** Ibid.

190 **On August 8, 1972, a forty-five-year-old Argentinian woman:** Casimiro Porras et al., "Recovery from Rabies in Man," *Annals of Internal Medicine* 85, no. 1 (1976): 44–48.

190 **One was a New York laboratory worker:** Centers for Disease Control, "Rabies in a Laboratory Worker," *Morbidity and Mortality Weekly Report* 26 (1977): 183–84.

190 **The second, a nine-year-old boy in Mexico:** Lucia Alvarez et al., "Partial Recovery from Rabies in a Nine-Year-Old Boy," *Pediatric Infectious Disease Journal* 13, no. 12 (1994): 1154–55.

190 **a six-year-old girl bitten by a street dog:** S. N. Madhusudana et al., "Partial Recovery from Rabies in a Six-Year-Old Girl," *International Journal of Infectious Diseases* 6, no. 1 (2002): 85–86.

192 **In a video made by her doctors:** Online resource accompanying Rodney E. Willoughby et al., "Survival After Treatment of Rabies with Induction of Coma," *New England Journal of Medicine* 352, no. 24 (2005): 2508–14.

192 **But by the time a second video was made:** Online resource accompanying William T. Hu et al., "Long-Term Follow-up After Treatment of Rabies by Induction of Coma," *New England Journal of Medicine* 357, no. 9 (2007): 945–46.

193 **In the spring of 2011, Giese graduated:** Mark Johnson, "Rabies Survivor Jeanna Giese Graduates from College," *Milwaukee Journal Sentinel,* May 8, 2011.

193 **On her YouTube channel, she has posted:** Jeanna Giese's YouTube channel, http://www.youtube.com/user/JeannaGieseRabies01#p/u.

193 **spelled out various unique features of Giese's case:** Willoughby et al., "Survival After Treatment of Rabies with Induction of Coma."

194 **On a Web site hosted by the Medical College of Wisconsin:** Children's Hospital of Wisconsin rabies registry home page, http://www.chw.org/display/PPF/DocID/33223/router.asp.

194 **In 2011, Precious Reynolds:** Stephen Magagnini, "Scrappy 8-Year-Old from Humboldt Beats All Odds in Her Battle Against Rabies," *Sacramento Bee,* June 13, 2011, 1B.

195 **Reynolds remained in a coma:** Erin Allday, "Rabies: Humboldt Girl Beats Virus Against Odds," *San Francisco Chronicle,* June 12, 2011, A1.

195 **Reynolds left UC Davis Children's Hospital:** http://sacramento.cbslocal.com/2011/06/22/girl-heads-home-after-surviving-rabies/.

195 **six out of thirty-five cases:** Ferris Jabr, "Rabies May Not Be the Invincible Killer We Thought," *New Scientist,* June 21, 2011.

196 **a thirty-three-year-old man treated at the King Chulalongkorn Memorial Hospital:** Thiravat Hemachudha et al., "Rabies," *Current Neurology and Neuroscience Reports* 6, no. 6 (2006): 460–68.

196 **Thiravat Hemachudha, was a vocal skeptic:** "The Girl Who Survived Rabies."

196 **in a subsequent paper, he and his colleagues:** Henry Wilde, Thiravat Hemachudha, and Alan C. Jackson, "Viewpoint: Management of Human Rabies," *Transactions of the Royal Society of Tropical Medicine and Hygiene* 102, no. 10 (2008): 979–82.

196 **Jackson penned a dissenting editorial:** Alan C. Jackson, "Recovery from Rabies," *New England Journal of Medicine* 352, no. 24 (2005): 2549–50.

196 **he remains unconvinced, and for an intriguing reason:** Alan C. Jackson, "Why Does the Prognosis Remain So Poor in Human Rabies?" *Expert Review of Anti-infective Therapy* 8, no. 6 (2010): 623–25.

197 **Pasteur himself recorded the case of a dog:** Hattwick et al., "Recovery from Rabies," quoting Louis Pasteur, Charles Chamberland, and Emile Roux, "Nouveaux faits pour servir à la connaissance de la rage," *Comptes Rendus de l'Académie des Sciences, Série III, Sciences de la Vie* 95 (1882): 1187–92.

197 **recovery from rabies has been documented:** Theodore C. Doege and Robert L. Northrop, "Evidence for Inapparent Rabies Infection," *Lancet,* Oct. 5, 1974, 826–29.

197 **One early nineteenth-century physician claimed in the *Lancet:*** "Preventive and Curative Treatment of Rabies," *Lancet*, September 29, 1838, 55–58.

197 **reported recovery from rabies after they transfused serum:** "Rabies," *Medical Annals of the District of Columbia* 33, April 1964: 158–59.

197 **nine cases of reported recovery:** Hattwick et al., "Recovery from Rabies."

197 **A survey for serum rabies antibodies:** Ibid.

197 **a case report that detailed an apparently unvaccinated survivor:** Centers for Disease Control and Prevention, "Presumptive Abortive Human Rabies—Texas, 2009," *Morbidity and Mortality Weekly Report* 59, no. 7 (2010): 185–90.

198 **At home, the girl's headaches resumed:** Ibid.

198 **Despite an extensive workup:** Ibid.

198 **The next day, the CDC ran tests:** Ibid.

199 **On March 14, the girl received a dose:** Ibid.

199 **"we need to focus more on prevention":** Barbara Juncosa, "Hope for Rabies Victims: Unorthodox Coma Therapy Shows Promise," *Scientific American,* Nov. 21, 2008.

CHAPTER 8: ISLAND OF THE MAD DOGS

203 **it was probably Thomas Aquino's dog:** Merritt Clifton, "How Not to Fight a Rabies Epidemic: A History in Bali," *Asian Biomedicine* 4, no. 4 (2010): 663–70.

203 **But enforcement of this law:** Ibid.

204 **Two months after Thomas's dog:** Ibid.

205 **When his mother took him to the hospital:** Luh De Suryani, "Rabies Threat Gets Ever More Real," *Jakarta Post,* Jan. 9, 2009.

205 **It took two more deaths:** Clifton, "How Not to Fight a Rabies Epidemic."

206 **in Kazakhstan:** International Society for Infectious Diseases, ProMED correspondence, July 19, 2011.

206 **dog bites are still responsible for:** The CDC's Rabies in the U.S. and Around the World page, http://www.cdc.gov/rabies/location/index.html.

206 **In South Africa's KwaZulu-Natal Province:** Chris Bateman, "AIDS Fuels Ownerless Feral Dog Populations," *South African Medical Journal* 95, no. 2 (2005): 78–79.

207 **But vaccination campaigns in dogs:** Partners for Rabies Prevention's introduction to the Blueprint for Rabies Prevention and Control, http://www.rabiesblueprint.com/spip.php?rubrique5.

207 **the cost of a full course of:** WHO Media Centre's Rabies Fact Sheet, http://www.who.int/mediacentre/factsheets/fs099/en/.

209 **the government had removed from targeted regions:** Desy Nurhayati, "Mass Culling of Stray Dogs to Continue Amid Protests," *Jakarta Post,* Nov. 6, 2009.

209 **one Australian woman described how her own dog:** "Bali Dog Cull Shocks Aussies," *Herald Sun,* March 1, 2009.

209 **Another Australian, a chef, witnessed:** Ibid.

210 **Krishna pointed out that it was in 1860:** http://bluecrossofindia.org/abc.html.

210 **But imported vaccines, which have been proven protective:** Luh De Suryani, "Vaccines Help Dogs Fight Rabies in Short-Term," *Jakarta Post,* Feb. 12, 2010.

211 **281 dogs had been destroyed:** "Hundreds of Dogs Put Down, Vaccinated Against Rabies in Bali," *Jakarta Post,* Dec. 18, 2008.

211 **its efforts to contain rabies on Bukit:** Luh De Suryani, "Denpasar Goes on Alert as More Rabid Dogs Found," *Jakarta Post,* Jan. 9, 2009.

211 **scores of high-ranking local government officials:** Luh De Suryani, "Rabies Death Toll Rises to Six," *Jakarta Post,* Jan. 19, 2009.

211 **despite the extermination of 26,705:** Desy Nurhayati, "Mass Culling of Stray Dogs to Continue Amid Protests," *Jakarta Post,* Nov. 6, 2009.

211 **Thomas Aquino's friend Freddy:** Luh De Suryani, "Rabies Death Toll Rises to Six."

211 **his three-year-old neighbor Ketut Tangkas:** Luh De Suryani, "Toddler Dies from Suspected Rabies," *Jakarta Post,* Jan. 6, 2009.

213 **immune dogs, or "warrior dogs":** Trisha Sertori, "Janice Girardi: Trusting in Warrior Dogs," *Jakarta Post,* March 29, 2010.

217 **Niels Pedersen, even gives some credence:** *Bali: Island of the Dogs* (2006).

217 **In addition to supplying owners with protection:** A. Agung Gde Putra, K. Gunata Dan, and Gde Asrama, "Dog Demography in Badung District the Province of Bali and Their Significance to Rabies Control" (paper presented at Konferensi Ilmiah Veteriner Nasional XI, Semarang, Central Java, Oct. 11–13, 2010).

221 **Bali's staggering canine turnover rate:** Ibid.

221 **abandoning the goal of eradicating rabies:** Luh De Suryani, "Administration Pushes Back Rabies-Free Deadline to 2015," *Jakarta Post,* May 26, 2011.

222 **"But Walt could see the drama":** Didier Ghez, ed., *Walt's People—Volume 10* (New York: Xlibris, 2010), 146.

223 **"there was no out but to kill the dog":** Ibid., 145.

223 **called the ending "child abuse":** C. Jerry Kutner, "Good Dog, Bad Dog: The Horror of *Old Yeller*," *Bright Lights Film Journal*, April 2001, http://www.brightlightsfilm.com/32/oldyeller.php.

223 **"still doesn't forgive us":** Didier Ghez, *Walt's People*, 145.

CONCLUSION: THE DEVIL, LEASHED

225 **In July 2009, though, a suspiciously erratic animal was spotted:** New York City Department of Health and Mental Hygiene, http://www.nyc.gov/html/doh/downloads/pdf/zoo/09vet07.pdf.

226 **raccoons thrive more today in urban:** Samuel I. Zeveloff, *Raccoons: A Natural History* (Washington, D.C.: Smithsonian Institution Press, 2002), 75.

226 **the raccoon eats everything:** Paul Rezendes, *Tracking and the Art of Seeing: How to Read Animal Tracks and Sign* (New York: HarperCollins, 1999), 162–63.

226 **"one of the most intensive wildlife rabies outbreaks":** Eugene Linden and Hannah Bloch, "Beware of Rabies," *Time*, Aug. 23, 1993.

226 **Before the mid-1970s, raccoon rabies:** Meghan E. Jones et al., "Environmental and Human Demographic Features Associated with Epizootic Raccoon Rabies in Maryland, Pennsylvania, and Virginia," *Journal of Wildlife Diseases* 39, no. 4 (2003): 869–74.

226 **more than thirty-five hundred raccoons:** Suzanne R. Jenkins and William G. Winkler, "Descriptive Epidemiology from an Epizootic of Raccoon Rabies in the Middle Atlantic States, 1982–1983," *American Journal of Epidemiology* 126, no. 3 (1987): 429–37.

226 **The mid-Atlantic saw its first case:** Victor F. Nettles et al., "Rabies in Translocated Raccoons," *American Journal of Public Health* 69, no. 6 (1979): 601–2.

227 **Within three decades, rabid raccoons:** Centers for Disease Control and Prevention, "Update: Raccoon Rabies Epizootic—United States and Canada, 1999," *Morbidity and Mortality Weekly Report* 49, no. 2 (2000): 31–35.

227 **updates to its online rabies map:** New York City Department of Health and Mental Hygiene Web site, http://www.nyc.gov/html/doh/downloads/pdf/cd/animal_rabies_2010_mn.pdf.

227 **rabid raccoons were staggering out:** Ibid., http://www.nyc.gov/html/doh/html/cd/cdrab-borough.shtml.

234–35 **Back in 1982, a Yale researcher named Thomas Lentz:** Thomas Lentz et al., "Is the Acetylcholine Receptor a Rabies Virus Receptor?" *Science* 215, no. 4529 (1982): 182–84.

235 **in a subsequent paper eight years later:** Thomas Lentz, "Rabies Virus Binding to an Acetylcholine Receptor α-subunit Peptide," *Journal of Molecular Recognition* 3, no. 2 (1990): 82–88.

235 **after treatment with the molecule, 80 percent:** Priti Kumar et al., "Transvascular Delivery of Small Interfering RNA to the Central Nervous System," *Nature* 448 (2007): 39–43.

235 **in March 2011, a team at Oxford:** BBC News, March 20, 2011.

236 **"Cerberus stood agape":** From Virgil's *Georgics.* See John Jackson trans., *Virgil* (Oxford: Clarendon Press, 1921), 101.

INDEX

Page numbers in *italics* refer to illustrations.